SUSTAINABLE ENERGY FOR ALL

Despite decades of effort and billions of dollars spent, two-thirds of people in Sub-Saharan Africa still lack access to electricity, a vital precursor to economic development and poverty reduction. Ambitious international policy commitments seek to address this, but scholarship has failed to keep pace with policy ambitions, lacking both the empirical basis and the theoretical perspective to inform such transformative policy aims.

Sustainable Energy for All aims to fill this gap. Through detailed historical analysis of the Kenyan solar PV market, the book demonstrates the value of a new theoretical perspective based on Socio-Technical Innovation System Building. Importantly, the book goes beyond a purely academic critique to detail exactly how a Socio-Technical Innovation System Building approach might be operationalized in practice, facilitating both a detailed plan for future comparative research as well as a clear agenda for policy and practice. These plans are based on a systemic perspective that is more fit for purpose to inform transformative policy ambitions like the UN's Sustainable Energy for All by 2030 initiative and to underpin pro-poor pathways in sustainable energy access.

This book will be of interest to academic researchers, policy-makers and practitioners in the field of sustainable energy access and low carbon development more broadly.

David Ockwell is Reader in Geography at the University of Sussex, UK, and Deputy Director of Research in the ESRC STEPS Centre. He is also a Senior Research Fellow in the Sussex Energy Group and a Fellow of the Tyndall Centre for Climate Change Research. David sits on the board of the Low Carbon Energy for Development Network.

Rob Byrne is Lecturer in SPRU (Science Policy Research Unit) at the University of Sussex, UK. With David, Rob co-convenes the Energy and Climate Research Domain of the ESRC STEPS Centre. He is also a Research Fellow in the Sussex Energy Group and a Fellow of the Tyndall Centre for Climate Change Research. Rob sits on the board of the Low Carbon Energy for Development Network and is a member of Climate Strategies.

PATHWAYS TO SUSTAINABILITY

This book series addresses core challenges around linking science and technology and environmental sustainability with poverty reduction and social justice. It is based on the work of the Social, Technological and Environmental Pathways to Sustainability (STEPS) Centre, a major investment of the UK Economic and Social Research Council (ESRC). The STEPS Centre brings together researchers at the Institute of Development Studies (IDS) and SPRU (Science Policy Research Unit) at the University of Sussex with a set of partner institutions in Africa, Asia and Latin America.

Series Editors:
Ian Scoones and Andy Stirling
STEPS Centre at the University of Sussex

Editorial Advisory Board:
Steve Bass, Wiebe E. Bijker, Victor Galaz, Wenzel Geissler, Katherine Homewood, Sheila Jasanoff, Melissa Leach, Colin McInnes, Suman Sahai, Andrew Scott

Titles in this series include:

Transforming Health Markets in Asia and Africa
Improving quality and access for the poor
Edited by Gerald Bloom, Barun Kanjilal, Henry Lucas and David H. Peters

Pastoralism and Development in Africa
Dynamic change at the margins
Edited by Ian Scoones, Andy Catley and Jeremy Lind

The Politics of Green Transformations
Ian Scoones, Melissa Leach and Peter Newell

Governing Agricultural Sustainability
Global lessons from GM crops
Phil Macnaghten and Susana Carro-Ripalda

Gender Equality and Sustainable Development
Edited by Melissa Leach

Adapting to Climate Uncertainty in African Agriculture
Narratives and knowledge politics
Stephen Whitfield

One Health
Science, politics and zoonotic disease in Africa
Edited by Kevin Bardosh

'This book makes a hugely valuable, timely and action-oriented contribution to academic and policy debates about sustainable energy access in what is without doubt the most critical and insightful treatment of the subject to date. Bringing together hitherto unconnected fields of scholarship it moves well beyond the usual preoccupations with the technical and economic to productively explore some of the neglected historical, socio-cultural and political dimensions of energy access. Combining comparative empirical analysis with a very sophisticated and insightful conceptual framework concerned with national socio-technical innovation systems this book is essential reading for anyone seeking a more systematic understanding both of the specific challenges of enhancing sustainable energy access and of wider pro-poor green transformations.'

— *Marcus Power, Professor of Geography, Durham Energy Institute, University of Durham, UK*

'This is a highly significant book and should be read by anybody interested in the challenges surrounding the pursuit of expanded access to modern energy services and low carbon transitions in the so-called developing world. Eschewing the dominant foci on hardware financing and private sector entrepreneurship, Dave Ockwell and Rob Byrne call for a much more nuanced and systemic understanding of how transformative change might be achieved. They emphasize in particular that sustainable energy access must be seen as an explicitly political problem "with solutions that are themselves political as much as they are financial, technical or social." I am one hundred percent in agreement with them.'

— *Ed Brown, Co-Chair of the Low Carbon Energy for Development Network and Senior Lecturer in Geography, University of Loughborough, UK*

"The book is a must read for all who are interested in the transitions to low carbon economies, especially in the energy sector. The Socio-Technical Innovation System Building proposed provides an alternative theoretical perspective for understanding the dynamics of emerging clean energy technologies in emerging countries."

— *Kevin Urama, Senior Advisor to the President on Inclusive and Green Growth, African Development Bank*

'This book is a thoughtful, timely, and very useful addition to the literature pertaining to the transfer, development, and adoption of sustainable energy technologies (and, indeed, other technologies with similar attributes), which in turn is critical to the attainment of climate and other sustainable development goals. Drawing on detailed empirical research on the solar PV market in Kenya as well as the literature on innovation systems and socio-technical transitions, Ockwell and Byrne's analysis will make a significant contribution both to the theoretical discussions and the growing practical efforts to effectively deliver sustainable energy technologies for the poor.'

— *Ambuj Sagar, Vipula and Mahesh Chaturvedi, Professor of Policy Studies, Indian Institute of Technology, Delhi, India*

'Translating the rhetoric of "sustainable energy access for all" into reality requires that we understand the political and economic work involved in the process of innovation and the reasons for its success and failure. *Sustainable Energy for All* gives us a new and unique insight into these processes as they are unfolding in Kenya and beyond, giving us the insights needed to understand how to realise the potential of energy transitions in Africa.'

— *Harriet Bulkeley, Professor of Geography, Durham Energy Institute, University of Durham, UK*

'The challenge of progressing towards a pro-poor, green and sustainable energy pathway in Africa is a pressing development question today. And yet, we lack the conceptual tools and empirical analysis that support socially informed policies for transformative change. Ockwell and Byrne have done us a tremendous service in meticulously unpacking the interdependencies between the social context and innovation opportunities for energy access in Africa. This is a must-read for anyone concerned with the issues of the *African energy project* – written with clarity and persuasive logic.'

— *Yacob Mulugetta, Professor of Energy and Development Policy, Department of Science, Technology, Engineering & Public Policy (STEaPP), University College London, UK*

SUSTAINABLE ENERGY FOR ALL

Innovation, technology and pro-poor green transformations

David Ockwell and Rob Byrne

LONDON AND NEW YORK

First published 2017
by Routledge
2 Park Square, Milton Park, Abingdon, Oxon OX14 4RN

and by Routledge
711 Third Avenue, New York, NY 10017

Routledge is an imprint of the Taylor & Francis Group, an informa business

British Library Cataloguing in Publication Data
A catalogue record for this book is available from the British Library.

Library of Congress Cataloging in Publication Data
Names: Ockwell, David G., author. | Byrne, Rob, author.
Title: Sustainable energy for all : innovation, technology and pro-poor green transformations / David Ockwell and Rob Byrne.
Description: Abingdon, Oxon ; New York, NY : Routledge is an imprint of the Taylor & Francis Group, an informa business, [2016] | Series: Pathways to sustainability | Includes bibliographical references.
Identifiers: LCCN 2016001141| ISBN 9781138656925 (hb) |
ISBN 9781138656932 (pb) | ISBN 9781315621623 (ebk)
Subjects: LCSH: Renewable energy sources--Africa, Sub-Saharan. | Clean energy industries--Africa, Sub-Saharan. | Energy security--Africa, Sub-Saharan. | Solar power--Kenya. | Sustainable Energy for All (Vienna, Austria)
Classification: LCC TJ807.9.A357 O45 2016 | DDC 621.0420967--dc23
LC record available at http://lccn.loc.gov/2016001141

ISBN: 978-1-138-65692-5 (hbk)
ISBN: 978-1-138-65693-2 (pbk)
ISBN: 978-1-315-62162-3 (ebk)

Typeset in Bembo
by Taylor & Francis Books
Printed in Great Britain by
Ashford Colour Press Ltd, Gosport, Hants

For Chloe, for the most important context of all, and for the Byrne and Knight families, for unshakeable support and lots of fun.

CONTENTS

ILLUSTRATIONS

Figures

Tables

PREFACE

We finished writing this book at the end of 2015, a year in which the international community adopted both the Sustainable Development Goals (SDGs) and – under the UN Framework Convention on Climate Change – the Paris Agreement. Both of these landmark decisions are relevant to the content of this book. The Paris Agreement, while imperfect, offers new political hope that international efforts to combat climate change can at last turn to action, following decades of struggle to reach consensus on a work plan. A significant element of the effort will be the achievement of climate change mitigation – and, potentially, adaptation – through the global deployment of low carbon technologies, and a large part of this will involve the building of low carbon energy systems in developing countries. The 17 SDGs, as with the Paris Agreement, are applicable globally and are meant to help international policy interventions promote sustainable development. One of the goals takes its inspiration from the UN's initiative, Sustainable Energy for All (SE4All), which sets the ambitious target – among others – of achieving universal access to energy by 2030.

As the title of this book makes clear, the energy goal is the specific focus of our attention here, and it is to achieve this goal that we have attempted to outline a theoretical framework we hope can inform much more effective policy interventions. But, in writing the book, it became clear to us that our theoretical framework, and its policy implications, may be applicable to sustainability more generally and so we have also attempted to highlight this where relevant. Of course, what we outline in this book can only be a contribution. We hope others will find inspiration in the ideas we articulate here and, to this end, we finish the book with some thoughts on a research agenda that could further develop and refine the theory in ways that support sustainable development, whether in the field of energy or otherwise.

ACKNOWLEDGEMENTS

The book reflects the work of many years for both the authors. Along the way, we have had the good fortune to be involved in many stimulating discussions, debates, deliberations, workshops, walkshops, reading groups, retreats, lectures, seminars, pub nights and parties. Keeping us company – and on our toes – have been friends and colleagues, critics and collaborators, and mentors, students and supervisors. As a result, it is impossible to point to everyone who has influenced our thinking. There are so many. Consequently, we can only offer a gesture of thanks to all of those who are not named below and hope they can forgive our frail memories.

We would like to acknowledge the influence and support of the many colleagues with whom we have worked, both in the STEPS Centre and at the University of Sussex as well as others across the world. But, specifically, we wish to thank the following: Saurabh Arora, Lucy Baker, Andrew Barnett, Sarah Becker, Martin Bell, Ed Brown, Rose Cairns, Roberto Camerani, Ben Campbell, Jon Cloke, Julia Day, Heleen de Coninck, Yusuf Dirie, Harriet Dudley, Adrian Ely, Tim Forsyth, Lorenz Gollwitzer, Rüdiger Haum, Merylyn Hedger, Sabine Hielscher, Oliver Johnson, Phil Johnstone, Lukas Kariongi, Edith Kirumba, Melissa Leach, Gordon MacKerron, Alexandra Mallett, Fiona Marshall, Mipsie Marshall, Charles Memusi, Erik Millstone, Pete Newell, Cosmas Ochieng, Jose Opazo, Nathan Oxley, Nicholas Ozor, Prosanto Pal, Jon Phillips, Sandra Pointel, Marcus Power, Rob Raven, Paula Rolffs, Ambuj Sagar, Seela Sainyeye, Martin Saning'o, Estomih Sawe, Ian Scoones, Girish Sethi, Adrian Smith, Andy Stirling, Michele Stua, Blanche Ting, Kevin Urama, Anne-Marie Verbeken, Jim Watson, Bettina Zenz and two anonymous reviewers.

Apart from these acknowledgements, we would like to express our thanks to all those who agreed to be interviewed during the field research, and all those who assisted in many other ways. For the first period of Kenya fieldwork, Rob Byrne was hosted in Nairobi by Energy for Sustainable Development Africa (ESDA, now Camco Advisory Services). He is grateful to Stephen Mutimba and his staff for all their help. The second period of Kenya fieldwork involved a project partnership with the African Technology Policy Studies Network, also in Nairobi. Again, Rob Byrne is grateful for all their assistance, especially Kevin Urama, Nicholas Ozor and Carol Thuku. And particular thanks are due to Andrew Kilonzo, Mike Harries, John Ng'anga Kimani, the staff at Ubbink East Africa, and Mark Hankins. Rob is especially grateful to Mark Hankins' family who cheerfully suffered an entire weekend at their home outside Nairobi during which Mark and Rob did nothing but talk about solar. They deserve as much sympathy as gratitude.

Finally, we would like to gratefully acknowledge the various funders who have made the research and writing possible. They include the ESRC (grant numbers PTA-031-2004-00227 and ES/I021620/1), the UK Department of Energy and Climate Change, and the Climate and Development Knowledge Network.

ABBREVIATIONS

ABM	Associated Battery Manufacturers
ADB	Asian Development Bank
ADF	African Development Foundation
AEEP	Africa-EU Energy Partnership
AIBM	Automotive and Industrial Battery Manufacturers
APSO	Agency for Personal Services Overseas
AT	appropriate technology
BOP	Bottom of the Pyramid
BOS	balance of system
CBK	Cooperative Bank of Kenya
CCS	carbon capture and storage
CDM	Clean Development Mechanism
CGIAR	Collaborative Group for International Agricultural Research
CIC	Climate Innovation Centre
CPI	consumer price index
CRIBs	Climate-Relevant Innovation System Builders
CSC	Commonwealth Science Council
CTCN	Climate Technology Centre and Network
CTP	Climate Technology Programme
DC	direct current
DFID	Department for International Development
EAA	Energy Alternatives Africa
EBRD	European Bank for Reconstruction and Development
EPO	European Patent Office
ERC	Energy Regulatory Commission
ESD	Energy for Sustainable Development
ESDA	Energy for Sustainable Development Africa

ESMAP	Energy Sector Management Assistance Programme
FIT	feed-in tariff
GDP	gross domestic product
GEF	Global Environment Facility
GHG	greenhouse gas
GTZ	German Development Cooperation (Deutsche Gesellschaft für technische Zusammenarbeit)
IADB	Inter-American Development Bank
ICT	information and communications technology
ICTSD	International Centre for Trade and Sustainable Development
IFC	International Finance Corporation
IGAD	Intergovernmental Authority on Development
IGO	intergovernmental organisation
IMF	International Monetary Fund
IPR	intellectual property right
JICA	Japanese International Cooperation Agency
JKUAT	Jomo Kenyatta University of Agriculture and Technology
KARADEA	Karagwe Development Association
KEBS	Kenya Bureau of Standards
KENGO	Kenya Environmental Non-Governmental Organizations
KEREA	Kenya Renewable Energy Association
KES	Kenya Shilling
KESTA	Kenya Solar Technician Association
KSTF	KARADEA Solar Training Facility
LED	light-emitting diode
MERD	Ministry of Energy and Regional Development
MFI	micro-finance institution
MLP	multi-level perspective
MOE	Ministry of Energy
MW	megawatt
NASA	North American Space Agency
NDE	National Designated Entity
NGO	non-governmental organisation
NIS	National Innovation System
NITA	National Industrial Training Authority
OECD	Organisation for Economic Co-operation and Development
PAYG	Pay-As-You-Go
PV	photovoltaic
PVGAP	Photovoltaic General Approval Program
PVMTI	Photovoltaic Market Transformation Initiative
R&D	research and development
SACCO	Savings and Credit Cooperative
SDG	sustainable development goal
SE4All	Sustainable Energy for All

SEA-RIBs	Sustainable Energy Access Relevant Innovation-system Builders
SELF	Solar Electric Light Fund
SHS	solar home system
SME	small and medium-sized enterprise
SNM	Strategic Niche Management
SPL	solar portable lantern
SSA	Sub-Saharan Africa
STEP	Solar Technician Evaluation Project
STI	science, technology and innovation
STISA	Science, Technology and Innovation Strategy for Africa
TEC	Technology Executive Committee
UNDP	UN Development Programme
UNEP	UN Environment Programme
UNFCCC	UN Framework Convention on Climate Change
USAID	United States Agency for International Development
WB	World Bank
WCED	World Commission on Environment and Development
WHO-EPI	World Health Organization Expanded Programme on Immunization

1

INTRODUCTION

Beyond hardware financing and private sector entrepreneurship

Low carbon Africa?

An unimaginable number of people on the planet today lack access to electricity, something that is now fundamental to many aspects of human and economic development. For some, it is so important that we 'might even ... consider access to electricity as a human right' (Winther 2008, p. 224). Globally, the number of people lacking electricity access sits at 1.1 billion (SE4All 2015, p. 2). In Africa, this translates to an average of two out of every three people, but this disguises huge variations across and within countries. In Kenya, the focus of much of this book, for example, the figure rises to almost four in every five people lacking access to electricity, and more than nine in ten in rural areas (SE4All 2015, p. 41). Nevertheless, examination of Kenya's highly dynamic market in off-grid photovoltaic technologies (hereafter, solar PV) suggests ways in which to significantly improve these electricity access numbers and, hopefully, the prospects for human and economic development. Based on detailed empirical analysis of this promising Kenyan example – and supported by examples from research in Tanzania, India and China – it is the aim of this book to introduce a systemic conceptual framework through which to understand how research, policy and practice can provide more effective analyses and interventions to address the electricity access problem.

The statistics quoted above are abstractions that perhaps make it difficult to comprehend the enormous impact on everyday life they are meant to represent. Instead, for those of us living in the rich world or those who have long had access to plentiful electricity, it may be more helpful to reflect on how electricity is involved in our daily routines. Stop for a moment and look around you. Most likely, everywhere you look there will be electrical appliances (you may even be reading this on one). Think through your average day, from getting up in the morning through to going to bed at night, and note every time you rely on electricity: from boiling the kettle, to washing and ironing clothes, to lighting and heating your home, or simply turning on a television or radio, or charging a mobile phone. So many aspects of our lives, many of them basic human needs – lighting, heating, cooling, cooking, washing and communication – are made easier or, indeed, possible because of our access to electricity. Furthermore, many of the goods and services we consume, and many of the jobs we do to earn money, are also only possible because we have access to reliable electricity.

No wonder then, in the year 2015, such a stark difference between the lives of the world's rich and the world's poor has driven ambitious policy commitments to try to rectify the issue of electricity access. In 2011, under the leadership of Ban Ki-moon, the United Nations (UN) announced a commitment to providing 'sustainable energy for all' (SE4All) by 2030. Note the inclusion here of 'sustainable' energy, connoting the nexus between energy access and climate change, and environmental sustainability more broadly. It also raises the possibility of using renewable energy sources, often off-grid, to provide electricity to many of the people currently lacking access; certainly for those who live in rural areas where grid extension is prohibitively expensive, but also for those in slum urban areas where expense prevents connection to the existing electricity grid.

Africa's[1] economy and accompanying energy demands have almost doubled in size since the turn of the century and it is estimated it will see further increases in energy demand of up to 80 per cent by 2030 (IEA 2014). If initiatives such as SE4All succeed in getting large numbers of renewable energy technologies into use, then the prospect of Sub-Saharan Africa (SSA) locking into lower carbon development trajectories is a powerful one, although we should note that this is not without controversy. After all, most SSA countries are already 'low carbon' from a per capita or aggregate greenhouse gas (GHG) emissions perspective. Considering that the energy needs of the poor are (currently) small, some analysts and practitioners argue that the poor should not be constrained to using low carbon technologies, as their GHG emissions will not significantly increase the global total even if they were to meet all their energy needs with fossil energy sources (e.g. see Sanchez 2010). This argument is linked to questions regarding the extent to which renewable energy technologies, particularly solar PV, can support economically productive activities, or anything beyond basic services such as lighting, mobile phone charging and social connectivity through television and radio.

On the face of it, these two points present challenges to the argument for promoting pro-poor low carbon development. But there are counter-arguments. First, while the poor may be surviving on small quantities of energy at present, projections that they will not increase their energy consumption much into the future could merely reflect limited ambition – or contestable modelling assumptions – on the part of analysts (e.g. see Bazilian and Pielke 2013). Building on this observation, Bazilian and Pielke (2013, p. 75) caution:

> The lower the assumed scale of the challenge, the more likely it is that the focus will turn to incremental change that amounts to 'poverty management,'

rather than the transformational changes that will be necessary if we are to help billions climb out of poverty.

In other words, the point of addressing energy access is to enable the poor to escape poverty, not to be a little less poor. As they become wealthier, we can expect them to increase their energy consumption and, as Wolfram et al. (2012) argue, this increase could be highly significant over the long term. In the meantime, if there are no carbon constraints, the establishment of the supporting energy infrastructure, social and technical practices, political and economic interests, sunk investments, laws and regulations, and so on, associated with fossil-based provision of energy would mean the poor becoming locked into high carbon development pathways (see Unruh 2000 for an explanation of the lock-in idea). That is, promoting fossil-based energy access would be promoting development pathways that just store up problems for developing countries that they will have to address later.

The second point, which questions whether renewable energy technologies can support economically productive activities, is in some ways more difficult to challenge. Technically, there are few reasons why renewable energy technologies could not support the entire range of productive activities. Such activities, as we implied earlier, require energy in the form of electricity, or heat, mechanical power, etc. (Modi et al. 2005). But electricity generated from a solar PV module is still electricity; heat generated from burning biogas is heat; mechanical power generated from a windmill is mechanical power, and so on. When it comes to renewable energy technologies, the main technical challenge for supporting productive activities is not so much the kind of energy generated by a specific technology; it is, instead, about whether the energy can be delivered fast enough for the activity in question. That is, the challenge is whether the specific technology can generate enough power, and whether this power can be maintained as needed or whether there is an intermittency issue. The other main 'technical' challenge is whether the cost of generating power from a specific technology is cheap enough to ensure that productive activities are economically viable, especially when compared with other available options.

The power, intermittency and cost of renewable energy technologies are all dynamic characteristics rather than fixed quantities, and they are changing in favourable ways. Both power and intermittency issues could be addressed through energy storage and management technologies, such as better batteries and 'smart' grids, or a combination of both. While there is still a long way to go in this regard, an interesting development in battery technology – batteries that are designed to work on the grid as well as off-grid – was announced by the firm Tesla[2] in 2015, but there is also plenty of other research into batteries that could yield important benefits (e.g. see Van Noorden 2014). And the evidence of favourable changes in the cost of renewable energy technologies is now strong and clear. For example, the costs of generating grid-connected electricity from renewable energies are falling rapidly and are already competitive with fossil fuel options, even after accounting for the costs of addressing intermittency (IRENA 2015).

These favourable changes in technical characteristics provide some of the reasons why the global deployment of renewable energy technologies has been accelerating. According to REN21 (2015, p. 17), in 2014, there were more additions of renewables to global power capacity than coal and gas combined, and renewables were able to supply almost a quarter of global electricity. Although these increases are not yet happening fast enough to meet the goals of policy initiatives such as SE4All, these kinds of changes are inspiring some analysts to investigate the feasibility of a rapid and complete worldwide replacement of fossil-based energy systems with renewables. One example is the work done by Mark Jacobson at Stanford University and Mark Delucchi at the University of California who, together, have published peer-reviewed work modelling the feasibility of providing energy for all global purposes by 2030 using only water, wind and solar power (see Delucchi and Jacobson 2011; and Jacobson and Delucchi 2011). Although their modelling has been critiqued (see Trainer 2012), they have strongly defended both it and their findings (Delucchi and Jacobson 2012).

But, returning to the policy commitment of sustainable energy for all, we can further interrogate the word 'sustainable' in relation to another aspect, going beyond the technical or physical that a focus on environmental sustainability privileges. That is, we can think about it in its fullest sense, drawing on the widely used definition of sustainable development as first articulated in the World Commission on Environment and Development (WCED) report, *Our Common Future*. The familiar definition given in the report is, of course, 'Sustainable development is development that meets the needs of the present without compromising the ability of future generations to meet their own needs' (WCED 1987, p. 43). The report then expands on this definition and, in particular, emphasises that sustainability is not just about the environment. On the same page, it goes on to say:

> Development involves a progressive transformation of economy and society. A development path that is sustainable in a physical sense could theoretically be pursued even in a rigid social and political setting. But physical sustainability cannot be secured unless development policies pay attention to such considerations as changes in access to resources and in their distribution of costs and benefits. Even the narrow notion of physical sustainability implies a concern for social equity between generations, a concern that must logically be extended to equity within each generation.

There are deeply political implications arising from this elaborated definition, not least of which is the concern for social equity. The expression used may be timid – 'concern for social equity' rather than, say, 'commitment to achieving social equality' – but it nevertheless points to an essential characteristic of sustainability: that development will not be sustainable if it ignores – or worsens – social justice outcomes. It follows, then, that a commitment to 'sustainable' energy for all must incorporate not just a commitment to environmentally and economically sustainable energy but also a commitment to its social dimensions as well. This has implications for the kinds of interventions that policies might drive. But it also has implications for the ways in which we might understand, analyse and recommend interventions, all of which arise to a large extent – although not exclusively – from academic debate.

Sustainable energy access and the scholarly deficit

We wrote this book in 2015 and the numbers above give some idea of the level of ambition that policy commitments like SE4All imply. 'Transformation' is an overused word in academic discussions on issues of sustainability these days, but providing sustainable electricity access to more than one billion people over the next 15 years (the UN's 2030 target) implies nothing less than a transformation. The notion of 'transformation' is understood here to mean change that is both rapid and wide-reaching, in terms of the number of additional poor people gaining access to sustainable energy, but also change that works for social justice. Echoing the WCED sustainable development definition, we could accept the possibility of all poor people getting access to economically and environmentally sustainable energy (cf. physical sustainability) while achieving minimal social justice outcomes. For example, we could imagine a scenario in which every off-grid household gets a solar PV system without having any transformative impact on gendered power relations regarding intra-household access to clean lighting services (see Jacobson 2004 for some evidence of unequal access to electricity in solar-powered households in Kenya). We might describe this as a shallow transformation.

In some ways, mainstream 'development' interventions of the kind traditionally associated with institutions such as the World Bank and the International Monetary Fund (IMF) could, in this regard, suffice to achieve the SE4All transformation. These interventions have been concerned with economic growth, defining their 'one-size-fits-all' policy prescriptions primarily from a neo-classical economics perspective. Critiqued by many, this kind of approach is blind to contexts and different views on what constitutes 'the good life', and subordinates the social to the logic of markets (e.g. see Escobar 2012 for perhaps the most elaborated critique of this approach). However, if we are serious about realising social equity, which we have argued above is essential to sustainability, then we must also work for fairer social relations – what we might call a deep transformation. Such a deep transformation might be catalysed by initially shallow transformative action – perhaps through technical improvements in access to energy that mean more households gain solar PV systems or grid connections – that enable deeper changes to happen over time.

But we cannot assume that these will happen automatically. Rather, achieving sustainable energy for all, in its fullest sense (including social justice and social equity), will require political work, not just technical action (Scoones et al. 2015b) at all levels from local to international and among powerful actors well beyond the SE4All initiative. For example, sustainable energy access also forms a core pillar of efforts under the Africa-EU Energy Partnership (AEEP n.d.); the African

Development Bank's 2013–2022 strategy is predicated on driving industrialisation across Africa through Green Growth, maximising opportunities for low carbon energy technology markets (AfDB 2013); multiple international donors have reframed their strategic approaches around widely used, but ill-defined, concepts such as 'green growth', 'low carbon development' and 'climate-compatible development' (see Mulugetta and Urban 2010 for a discussion of the various interpretations of low carbon development); and, at the level of international climate policy negotiations under the UN Framework Convention on Climate Change (UNFCCC), the transfer of low carbon energy technologies to developing countries remains central to achieving both GHG emissions reductions and national development goals (UNFCCC 2015).

Importantly, these policy ambitions implicitly assume that such a transformation can be driven by the deliberate interventions of key actors. These actors could be individuals or organisations, intervening through policy and practice at a range of possible scales, from international to local. But deliberate interventions to achieve transformative change in energy access are unprecedented. Doing this with low carbon energy technologies is even more challenging, given that they are marginalised (economically, politically and socially) relative to high carbon energy technologies. Furthermore, much of Africa lacks the infrastructure, technological capabilities and functioning innovation systems for even these mainstream technologies. But, while this presents an unprecedented challenge (and notwithstanding the potential controversy noted above), it also represents an unprecedented opportunity. Sub-Saharan Africa's lack of existing infrastructure offers the region, more than any other, the potential to develop along lower carbon pathways, rather than locking-in (Unruh 2000) to the high carbon, fossil fuel-based infrastructure that is now so difficult for other nations and continents to decarbonise in the bid to tackle climate change.

Myriad actors (individuals and institutions) are currently operating across Africa trying to drive sustainable energy access; from small, local non-governmental organisations (NGOs), to national governments, international donors, multinational companies, regional governmental organisations and multi-billion dollar programmes coordinated by intergovernmental organisations such as the UN and the World Bank. But what do interventions that transform sustainable energy access in low-income countries look like? Why have so many past interventions failed to drive change on a wider scale or at a more rapid rate? In the handful of examples where transformative changes in sustainable energy access have occurred, what drove them? What made them transformative as opposed to narrow, slow and incremental? Who was involved? What did they do?

These are all questions with which we seek to engage in this book. We do not pretend to be able to answer all of them. We do argue, however, that the conceptual framework we develop in this book is better equipped to inform the transformative policy ambitions mentioned above than the two-dimensional approach of the majority of existing literature on energy access in Sub-Saharan Africa. That literature is dominated by a focus on finance and technological hardware and an accompanying dominance of economics and engineering-based analyses

(Watson *et al.* 2012). Notwithstanding a handful of recent contributions (e.g. Jacobson 2007; van Eijck and Romijn 2008; Romijn and Caniëls 2011; Sovacool and Drupady 2012; Byrne 2013b; Baker *et al.* 2014; Rolffs *et al.* 2015; Naess *et al.* 2015), there are few academic contributions that go beyond technical and economic analyses, and almost none that consider the socio-cultural and political dimensions of energy access in SSA. A recent systematic review demonstrated that the literature is characterised by a range of disparate and uncoordinated efforts. Studies consist of project-by-project, or policy-by-policy, analyses of 'barriers' and few are of high enough quality to contribute to more systematic learning (Watson *et al.* 2012). In many of these analyses, there is an increasing use of the term 'enabling environment' to describe the context that facilitates change. This tends to be a catch-all term for anything beyond financial or technical challenges. There is little in the way of any comprehensive articulation or explicit theorising of what constitutes such enabling environments.

Therefore, the argument we make in this book begins with the observation that scholarship has not kept up with policy ambitions in this field. The existing literature lacks both the necessary conceptual tools and the empirical basis to answer the questions posed above and to inform policy approaches that might be fit for purpose in realising current transformative ambitions. Only a handful of real-world, empirical examples exist that look anything like a transformation in access to sustainable energy technologies for poor people in Africa. No attempt has yet been made to systematically analyse these examples and act upon the lessons that might be learned. Doing so requires new conceptual thinking and comparative empirical analysis that bridges traditional boundaries between hitherto unconnected fields of scholarship. Extending from this, we also need an action-oriented focus that is able to both advance scholarly understandings and inform contemporary policy and practice. It is this lacuna that we seek to address in this book by introducing a systemic analytic perspective based on empirical analyses of the solar PV market in Kenya – one of the few examples of more transformative change in sustainable energy access that do exist in Africa. Of course, this one empirical example could easily be dismissed as an insufficient basis upon which to construct a conceptual framework. This is debatable but, nevertheless, we seek to address this criticism by supporting the insights this one example offers with those from low carbon research in Tanzania, India and China, as well as the innovation studies literature more generally.

The problem with hardware financing and private sector entrepreneurship

Scholarly work in the climate policy literature, particularly drawing on environmental economics, offers a more generic perspective on the problem of low carbon energy technologies and developing countries than is seen in the more specific literature on energy access in Sub-Saharan Africa. The generic perspective is one that has gained significant international policy traction. In this literature, the problem is theorised as the standard economic explanation of market failure in relation to the transfer of low carbon technologies[3] to developing countries. Little attempt is made to engage with ideas of how poor people might gain access to low carbon technologies, once they are present in a country. The implicit assumption is that transfer equates with access: fix the market and the energy access problem is solved.

The core focus of this environmental economics approach is the fact that markets for low carbon technologies do not capture the associated positive externality of reduced future carbon emissions. This means that there is no incentive for developing countries to invest in low carbon technologies, and hence a lack of developing country markets to attract investment from international technology-owning companies. Market-based policy mechanisms are therefore prescribed to meet the incrementally higher costs of low carbon energy technologies.

The classic example of the operationalisation of this perspective in practice is the Clean Development Mechanism (CDM), an instrument introduced under the Kyoto Protocol and hailed at the time as a 'win-win' agreement between the industrialised and developing countries that would promote both climate mitigation and sustainable development goals (Matsuo 2003; Lecocq and Ambrosi 2007). But, as Figure 1.1 shows, the CDM has led to an uneven distribution of investment, with Africa as a whole hardly benefiting at all (with just 3 per cent of cumulative

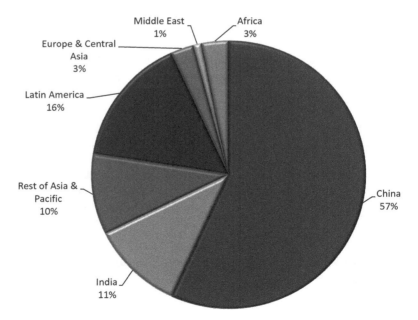

FIGURE 1.1 Distribution of cumulative investment under the CDM
Note: Figures are the percentage of total accumulated investment by the end of November 2015.
Source: Based on analysis of the CDM pipeline (www.cdmpipeline.org).

investment under the CDM to date). Essentially, the CDM can be characterised as a 'hardware financing mechanism' (Byrne *et al.* 2012b), emphasising its focus on finance for technology hardware. Other more targeted examples of hardware financing approaches also exist; in particular, efforts under some forms of intervention through the Global Environment Facility (GEF) that target low-income countries. However, these efforts have led to similar results. For example, as we discuss in more detail in Chapter 6, the GEF-financed Photovoltaic Market Transformation Initiative, implemented through the International Finance Corporation (IFC) and trialled in three countries, including Kenya, did not lead to 'market transformation'. Despite making USD 5 million in finance available, the initiative managed to add just 170 solar home systems (SHSs) to the market in Kenya (Byrne 2011, p. 129) – a market with annual sales of around 20,000–25,000 (Ondraczek 2013, p. 409). Clearly something is being missed by these 'hardware financing' approaches.

The other dominant approach that has emerged in this field more recently is a focus on private sector entrepreneurship. The classic example of this in relation to low carbon energy technologies is the establishment of Climate Innovation Centres (CICs) under the Climate Technology Programme directed by infoDev (the World Bank) with the UK Department for International Development (DFID) and their Danish counterparts (Danida). The CICs essentially operate as venture capital mechanisms, providing finance and business incubation support to early stage innovators in the countries where they operate (as of 2015, CICs are operational or under development in Kenya, Ethiopia, Ghana, South Africa, Morocco, the Caribbean, India and Vietnam) (infoDev 2015).

The idea of private sector entrepreneurs driving innovation and technological change in developing countries seems to have captured the imagination of international policy-makers and donors. It fits in neatly with normative commitments to neo-liberal ways of doing development. But it is ill-conceived for the specific circumstances that exist in a wide range of different contexts: differences in relation to types of technologies; differences in social practices facilitated by technologies; differences in socio-cultural variations of these practices; differences in levels of technological capabilities existing in different countries, regions or communities; differences in politics and political economies, and so on (Ockwell and Mallett 2012). Rather, it represents a renewed commitment to development approaches characterised by mono-economic assumptions, i.e. that economic 'laws' are applicable across time and space (Selwyn 2014, p. 8) and are not contingent on any of the context-specificities mentioned above. However, there is little evidence that financing early stage innovators is likely to translate into widespread systemic low carbon technical change in developing countries, let alone more transformative change that facilitates access to electricity for the world's poorest people.

As we discuss in Chapter 2, this is not surprising once we understand the insights gained from decades of research in the field of Innovation Studies, including recent contributions that focus specifically on low carbon energy technologies. This body of research, often focussed on the adoption and development of new technologies in developing countries, clearly demonstrates that widespread technological change occurs over time, via processes of developing technological capabilities and well-functioning innovation systems. It implies that the kind of private sector entrepreneurship supported by the CICs and others is likely to lead to a number of isolated businesses promoting new technological hardware, as opposed to making a contribution to systemic long-term change. Indeed, the academic paper by Sagar *et al.* (2009), upon which the CIC idea is based, pitches CICs as a means of strengthening and building innovation systems in developing countries, with activities that go well beyond the financing of entrepreneurship. The way in which CICs have been operationalised therefore falls well short of the original proposition made by Sagar *et al.*

Building innovation systems forms a core pillar of the conceptual framework that we develop in this book, albeit one that we argue needs to be extended beyond that introduced by Sagar *et al.* (2009). This extension needs to encompass a socio-technical understanding of change, and be embedded in more localised governance structures than are implied by Sagar *et al.*'s international focus. This is not to say that the CIC will not produce any pro-poor energy technologies or important pro-poor benefits. There is at least one example of the Kenyan CIC supporting a solar-powered irrigation technology that may well be of benefit to poor farmers. Rather, the point is that this kind of isolated investment is unlikely to result in widespread transformative change unless it is part of a more systemic approach to understanding – and seeking to effect – socio-technical change. It is this systemic approach to understanding and action that this book articulates.

As we argue in more detail in Chapter 2 and Chapter 3, these 'Hardware Financing' and 'Private Sector Entrepreneurship' framings, whether applied to the specific problem of sustainable energy access or the broader issue of low carbon technology transfer and development, miss a fundamental understanding of how technological change, technology adoption, innovation and development occur.[4] It is the aim of this book to develop a more sophisticated, systemic conceptual framework that is better able to explain examples of transformative change in sustainable energy access and, hence, inform future governance and interventions in practice. But, before summarising the details of this conceptual framework, let us first spend some time articulating why the current dominance of 'Hardware Financing' and 'Private Sector Entrepreneurship' framings of the problem matter so much.

A pathways perspective on why framings matter

We adopt the Pathways Approach (Leach *et al.* 2010a) as the normative starting point for our analysis in this book, building on its operationalisation in Byrne *et al.* (2014b) and Marshall *et al.* (in press). In simple terms, this approach casts aside the idea of a single, incontestable and normatively 'good' pathway of development, instead emphasising the need to remain open to multiple alternative development pathways that might be pursued. This is vital in the context of the complex, interrelated challenges resulting from the need to address poverty while simultaneously dealing with other (sometimes competing) priorities such as addressing climate change, environmental integrity, job creation, economic growth and social justice. It demands explicit recognition that there is no single, universally applicable, pathway towards achieving sustainable energy access, nor is there any single outcome or development trajectory that such pathways might unquestioningly support. Rather, multiple possible pathways and multiple potential 'destinations' exist, all of which have material consequences for the distribution of benefits that result along the way: who wins, who loses, whose interests are represented and whose are marginalised. The societal services and functions that sustainable energy technologies facilitate – such as providing light and connectivity to poor people in remote rural or slum urban settings, or the industrial energy needs of large businesses, etc. – are realised dynamically out of the interplay of various co-evolving complex systems (socio-cultural, technological, environmental, political, economic) and any particular unfolding of these dynamics constitutes a specific development pathway among multiple possibilities (Leach *et al.* 2010a).

Each of these complex systems themselves, and their combination, can be framed in different ways. Fundamentally, the Pathways Approach recognises that who you are shapes how you 'frame' – or understand – a problem or opportunity, and that this understanding tends to focus on a specific development pathway to the neglect of alternative perspectives. Or it might simply represent the received wisdom (Leach and Mearns 1996) of donors or government agencies, or other powerful actors, who fail to appreciate the realities of a problem from different perspectives, such as a farmer, shopkeeper or mother.

Each framing informs – and is informed by – a narrative that interprets the world in a particular way, reflecting and reinforcing the perspective of the narrator, justifying particular actions, strategies and interventions in order to achieve certain goals. As understood here, a narrative is used to 'suggest and justify particular kinds of action, strategy and intervention' Leach *et al.* (2010b, p. 371) and so a narrative attempts to enrol actors and their resources into particular ways to achieve development goals. If this enrolment is successful, then a particular direction of development is privileged, the result of which is an unfolding pathway co-evolving contingently and uncertainly in the interplay between these privileging forces and the various complex systems noted above.

As narratives orientate actors and resources towards particular goals, employing particular strategies, so a pathway of development evolves. All actors are operating with incomplete knowledge, and so any particular perspective underdetermines what might constitute material reality. The Pathways Approach therefore proposes that it is vital to create opportunities for multiple pathways to evolve in order to meet the priorities and needs of different groups. However, narratives that resonate with the perspectives of powerful actors – those who are able to mobilise sufficient resources to support their strategies – may become institutionalised, whereas other narratives, such as those of the already marginalised, may fail to materialise, thereby perpetuating unequal distributions of power. Furthermore, once certain narratives dominate policy, the framings of issues therein can serve to further exclude alternative framings and further marginalise those actors who promote these alternatives. In this way, policy narratives and associated problem framings have material consequences, influencing the extent to which particular identities and power relations are either reinforced or redressed.

Thus, multiple framings, narratives and pathways are possible. Different groups of actors will interpret the world in different ways, interpretations arising from their own experiences, situations, understandings, values and interests. Favouring certain framings over others, they will seek to promote narratives that would help to create their preferred development pathways. Some narratives will be more dominant than others, perhaps because they are promoted by powerful actors, and are likely to become manifested in interventions. Other narratives remain marginalised, perhaps because they are promoted by groups who are themselves marginalised or powerless (Byrne *et al.* 2012b, Forsyth 2008).

The ways in which problems are framed – in the case of this book, the problem of sustainable energy access and related issues of low carbon energy technology transfer, low carbon development, green growth, and so on – and the ways in which narratives are used to justify these framings – thus become a critical focus for analysis. Our argument is that the existing framings of the sustainable energy access problem (both in the majority of the academic literature and in dominant policy approaches) as one constituted and solvable through a focus on hardware financing and private sector entrepreneurship, are such that the needs of poor countries and the poor people therein are unlikely to ever be met. We argue for a fundamental reframing of the problem to one constituted by the need to build well-functioning, pro-poor socio-technical innovation systems in developing (and, particularly, low-income) countries. We base this systemic perspective on a synthesis of core aspects of Innovation Studies and Socio-Technical Transitions theory – a synthesis developed to a large extent from in-depth empirical analysis of the success of the solar PV market in Kenya, but bolstered by insights we have gained from related empirical work in Tanzania, India and China, along with reference to decades of work by many others in Innovation Studies. But, before summarising the main elements of this new theoretical framework, and the core arguments on which it is based, we should acknowledge that we too are engaged in framing and narrative construction.

Our normative position and the aims of this book

It would be remiss of us, having acknowledged the importance of framings, not to acknowledge our own normative positions in relation to both this field of enquiry and our professional and personal perspectives more broadly, both as researchers and as private individuals. Both of us are white and male academics based in a Northern university (the University of Sussex in the UK), having grown up and been educated in the Global North. And we have both spent extended periods of time living and working in various countries in the Global South. Byrne, in particular, has an intimate knowledge of sustainable energy access issues in East Africa, having worked as an engineer installing solar home systems in Tanzania. Ockwell originally trained in Economics and Ecology, and later Political

Science, working for several years as a policy consultant in the UK. His academic life emerged from an original approach that combined insights from Political Science with work in the natural sciences, culminating in a focus on the politics of scientific knowledge and its (mis)use in policy. Byrne originally trained in and practised engineering before retraining in the fields of Innovation Studies and Science and Technology Studies. Both have gravitated towards a more Science and Technology Studies-oriented perspective on science and development policy. Both continue to engage with, and act as consultants to, national and international policy-makers on climate change and development, maintaining a particular interest in sustainable energy access in Africa and working closely with development partners based in Nairobi. A normative commitment to pro-poor social justice and environmental sustainability is definitive of both authors' perspectives on their research and ideas pertaining to any kind of meaning in relation to life and human well-being more generally.

We should also acknowledge that the alternative framing of sustainable energy access that we espouse in this book is open to – indeed, welcomes – critique and questioning just as much as those framings we portray as being limited in their ability to serve the needs of poor people (i.e. the Hardware Financing and Private Sector Entrepreneurship framings we characterise above). Our hope is not so much that the alternative framing we articulate in this book be taken as a panacea for tackling the problem of sustainable energy access. Rather, it is that this alternative framing, with its systemic perspective and focus on the needs and practices of poor people, will provide a very different perspective on the governance of sustainable energy access and low carbon development more broadly, including related ideas such as green growth. This alternative framing is one that privileges a situated perspective, rooted in local institutions that fundamentally demand that democracy – in all its messy, unpredictable, subversive and creative possibilities – be foregrounded in the ways in which pathways to sustainable energy access and low carbon development are imagined, navigated and practised (Stirling 2014; 2015a).

In this way, our hope is that the politics of sustainable energy access are moved to the foreground of both analysis and practice. Far from sustainable energy access being a neutral concern that might be addressed by technocratic engineering and finance interventions, it becomes an explicitly political problem with solutions that are themselves political as much as they are financial, technical or social. Indeed, we expect that politics feature in the lived experiences of all actors involved in the field – even the supposedly 'neutral' engineers and economists – albeit in ways that those actors do not always explicitly acknowledge. The conceptual framework and approaches to policy and practice, and governance more broadly, proposed in this book are therefore intended to facilitate a more political and democratic approach to the sustainable energy access problem – one that is cognisant of the systemic nature of innovation and socio-technical change. Finance, engineering and private sector entrepreneurship each play a role in these dynamics but they do not, either in and of themselves or in combination, constitute the entirety of a socio-technical innovation system.

A socio-technical innovation systems perspective

Thus, our argument is that fundamental problems exist with the current framings of the problem of sustainable energy access, in both the majority of academic literature and in policy and practice. We seek to demonstrate through the empirical analysis in this book that neither the two-dimensional Economics–Engineering framing that dominates the academic literature on energy access in Sub-Saharan Africa nor the Hardware Financing–Private Sector Entrepreneurship framings that dominate policy (and much contemporary practice) can explain examples of transformative change in low carbon energy technology adoption in developing countries. Therefore, solutions based on these underdetermined framings are unlikely to meet the sustainable energy access needs of developing (particularly low-income) countries or the poor people therein. This, we argue, is primarily due to a failure to understand three key things:

1. The systemic nature of innovation and its role in relation to broader technological change, particularly regarding the adoption and development of new technologies in developing countries.
2. The socially situated and co-evolutionary nature of new technology adoption in specific social contexts, particularly in relation to new technologies that must compete with existing, dominant technologies. In the case of low carbon energy access, dominant technologies include – among others – kerosene lanterns, biomass-based cooking stoves, and diesel generators and batteries for electrical equipment (whether for domestic or productive use).
3. The role that key actors (individuals or organisations) might play in driving transformative socio-technical change by attending to issues 1 and 2. After all, any deliberate attempt to address the problem of sustainable energy access implicitly assumes that some kind of actor can and will intervene to achieve such an outcome. This third point is directly and inextricably linked to governance.

It should be noted that by attending to these three elements we are not arguing that technological hardware and finance are unimportant, nor are we saying there is no role for hardware financing and private sector entrepreneurship – they are and there is. Any interpretation of this book as an attempt to create a position that dismisses the importance of hardware, finance and entrepreneurship would fundamentally misunderstand the point we seek to make. Our argument is that these aspects of the energy access problematic need to be understood from a systemic perspective: they are component parts – not the sole constituents – of a broader system. Moreover, we argue, by ignoring other systemic, socially situated and political considerations, a narrow focus on hardware and finance will never lead to transformative change in sustainable energy access.

Insights from Innovation Studies

The first step the book makes towards a systemic conceptual framework centres around insights from the field of Innovation Studies and linked fields such as

Innovation Management (explored in more detail in Chapter 2). In particular, this builds on recent scholarly efforts to connect insights from Innovation Studies with issues of international climate policy and climate technology transfer (e.g. Sagar and Bloomberg New Energy Finance 2010; Ockwell and Mallett 2012; Hansen and Ockwell 2014; de Coninck and Puig 2015; Watson *et al.* 2015). This work has sought to move beyond the dominant Hardware Financing framing by wrestling insights from the Innovation Studies literature into a framework that can deal with the context of often less mature low carbon technologies (Ockwell *et al.* 2008), new patterns of technology flows (Lema and Lema 2013), including South–South (Brewer 2008), and the conditions of policy urgency that characterise climate change (as opposed to temporally neutral accounts of conventional technology transfer) (Ockwell and Mallett 2012). This literature is helpful in focussing attention on the insight that technology is essentially constituted by knowledge, with technological hardware representing the artefact of applied knowledge. It draws attention to how firms and industries develop their technological capabilities over time as they access new technologies, progressing from new productive capabilities 'up' to more complex innovative capabilities (e.g. see Hobday 1995a; Bell 1997; Bell 2012). This process leads to capabilities to manage and drive technological change, underpinning broader processes of (potentially low carbon) economic development and industrial change in developing countries. It is through the accumulation and advancement of these technological capabilities across firms and industries in different country contexts that technological change and economic development occur.

The literature in this field provides a further valuable insight, situating technological change in the context of countries' 'innovation systems', emphasising the network of actors (e.g. firms, universities, research institutes, government departments, NGOs) within which technological change occurs and the strength and nature of the relationships between them (Ockwell and Byrne 2015). Taking this insight seriously, we are offering an explanation of why hardware financing policy mechanisms like the CDM fail to deliver to low-income countries. Such supposedly technology- and country-neutral market mechanisms are likely to attract foreign investment in countries where internationally competitive technological capabilities already exist among domestic firms and industries to some extent, and where national systems of innovation are conducive to such investment. Therefore, these market mechanisms reinforce the comparative advantages of countries with well-developed capabilities in relation to low carbon technologies (such as China, India and Brazil), but fail to effect change in countries where existing technological capabilities and innovation systems are weak or absent.

But, despite the value of the recent Innovation Studies-inspired scholarship on international climate policy and low carbon technology transfer, these perspectives suffer from some critical limitations in being able to deal with sustainable energy access in Africa. First, they have principally been developed and applied in the context of OECD and Asian Tiger economies and, in the subsequent focus on climate technologies, rapidly emerging developing economies, particularly India and China (e.g. Ockwell *et al.* 2010a; Lema and Lema 2013; Hansen and Ockwell

2014; Watson *et al.* 2015). This makes 'traditional' Innovation Studies ill-equipped to deal with low-income country contexts that lack modern energy (or other) infrastructure in many areas, or basic levels of technological capabilities and innovation systems (this is discussed in more detail in relation to the Socio-Technical Transitions scholarship below). Second, and perhaps more importantly, Innovation Studies fails to engage with the crucial role of technology users and the social practices that co-evolve with technologies – particularly with access to new technologies that afford such transformative social, economic and political potential as those that facilitate electricity-based services. With a problem such as sustainable energy access in Africa, any useful conceptual framework must enable us to pay attention to the social practices of the poor people whom we hope will gain access to such technologies. To address this need, we look to the field of Socio-Technical Transitions.

Insights from Socio-Technical Transitions

The burgeoning field of Socio-Technical Transitions is far better equipped than traditional Innovation Studies to facilitate attention to technology users (Byrne 2011), focussing as it does on the co-evolutionary relationship between social practices, technology and innovation. Its interest in societal 'transitions' also looks like something closer to the notion of 'transformations' that pertain to this book's focus on sustainable energy. Societal transitions are understood in this literature to be society-wide changes from one set of stable social and technical configurations that perform a 'function' in that society to a new set that could perform that function more sustainably (e.g. see Geels and Schot 2007). A relevant example for us here is the global effort to 'transition' from a fossil fuel-dominated energy system to one based on renewable energies. As energy is fundamental to all processes, there are many 'societal functions' associated with energy systems, e.g. mobility, communications, entertainment, and so on. Each has its own social and technical configuration, and each configuration will vary across different contexts. So, for example, home entertainment could involve television, social media, conversation and music-making in various configurations with context-specific cultural practices, social norms and group or personal identities. In one context it may be normal practice to watch television during social gatherings; in another, watching television at such times may be considered insulting to one's guests. Whatever the specific social (taken as shorthand for social, cultural and political dimensions) and technical (together, socio-technical) configuration – and the co-evolution of the various social and technical elements – in any given time or place, it performs its function within a broader context that includes supply chains, government regulations, related functions, socio-technical configurations, and so on.

The Socio-Technical Transitions literature tries to incorporate these multiple dimensions and their interdependent dynamics into a coherent conceptual framework. As a result, it has much to say about how change occurs. It conceptualises existing society-wide socio-technical configurations as stable 'socio-technical regimes', understood as rules shared by actors in functional domains (e.g. shared knowledge base, belief systems, mission, strategic orientation, etc., within a particular society's transport system) (Geels 2004). In the case of energy, as we noted above, regimes would currently refer to fossil fuel-dominated energy production and consumption. But renewable energy alternatives exist, of course, and attempts are well underway to substitute them for fossil fuels. Nevertheless, renewable energy technologies are still relatively marginal compared to fossil fuel-based technologies and so the Socio-Technical Transitions literature conceptualises them as 'niche' technologies. In this case, the literature is interested in how to 'manage' such niches to effect widespread change (e.g. Kemp et al. 1998; Raven 2005). That is, it is interested in how fossil fuel-dominated regimes might change as a result of either successful management of niches of sustainable energy technologies to the extent that they compete with dominant fossil fuel-based regimes; or of landscape-level (the broader context) changes such as rising social and political demands for low carbon energy; or, addressed more recently in the literature, through the destabilisation of socio-technical regimes (Turnheim and Geels 2013). However, despite the promise of the Socio-Technical Transitions field, it too suffers from a number of critical limitations in its ability to deal with the sustainable energy access problem in Africa.

The first limitation, to a greater extent even than the Innovation Studies literature, is that the field has been developed using a rich range of historical case studies based on transitions in mostly post-war European contexts. An emerging strand of the literature is beginning to engage more explicitly with the contexts of developing countries, but this has mostly focussed to date on rapidly emerging economies, especially India (e.g. Berkhout et al. 2010) and South Africa (e.g. Baker et al. 2014; Swilling et al. 2015). Only a few peer-reviewed journal papers have attempted to deal explicitly with issues related to energy access in low-income countries (van Eijck and Romijn 2008; Ulsrud et al. 2011; Ahlborg and Sjöstedt 2015; Rolffs et al. 2015; Ulsrud et al. 2015;). This leaves the Socio-Technical Transitions literature wanting in going beyond what Furlong (2014) refers to as the 'modern infrastructure ideal'. It is unable to account for the stark differences between the well-established energy infrastructures in European and other Northern contexts, and the complete lack of established energy (or other) infrastructures in the majority of low-income countries across Africa. What infrastructure does exist in these contexts mostly serves a minority of urban elites. Access to grid-based electricity is not an imminent prospect for the majority of poor people in either the rapidly expanding urban fringes or the remote rural areas of most low-income countries.

Moreover, there are other challenges for transitions approaches when trying to apply them in the contexts of low-income countries. For example, based on rich empirical case studies in SSA countries, Keeley and Scoones (2003, p. 6) argue that attending to the historical relationships between science, local knowledges and political styles – influenced, as they are in Africa, by different (current and past) experiences with colonialism, post-independence development efforts and international science, technology and innovation – demonstrates 'multiple variegated and located forms of "modernity"' that defy universalist description and prognosis. This implies much more than simply the assertion that context matters; rather, it begs fundamental questions of Transitions theory and galvanises a call to extend the thinking to (particularly low-income) developing country contexts that resist easy categorisation within the usual niche–regime–landscape typology of Socio-Technical Transitions terminology. In relation to sustainable energy access, it begs a number of questions such as: What constitutes the regime of energy provision for poor people in Africa? Is reliance on wood fuel for heat and cooking, kerosene for lighting and occasional diesel generators for electricity enough to constitute what might be referred to as regimes with which niches of low carbon alternatives (e.g. solar home systems) have to compete and broader landscape dynamics intersect? Are these regimes established enough to constrain or enable the agency of actors and their potential to effect, or even drive, the widespread adoption of sustainable alternatives? Or are they more open, less stable and amenable to change in ways not yet properly considered in the Transitions literature – ways that might render both analytical purchase and insights for action in seeking to galvanise pro-poor, low carbon development pathways?

In this book, we develop the idea of 'socio-technical innovation systems', as opposed to simply 'innovation systems', allowing for the adoption of what we argue are the most promising strands (in relation to sustainable energy access in Africa) of both the Innovation Studies and Socio-Technical Transitions literatures described above. This goes beyond the limits of the Innovation Studies literature by attending explicitly to the role of technology users and the co-evolutionary nature of technological change, innovation and social practice. Critically, however, it allows us to test the limits of these literatures in the contexts of low-income countries and energy access for poor people. Concerns regarding space and spatio-cultural contingencies, including Geography-inspired critiques of Transitions scholarship (e.g. Lawhon and Murphy 2012), are thus explicitly introduced into the analysis.

For our purposes in this book, a final limitation of both the Transitions and Innovation Studies literatures is their failure to deal with the political nature of socio-technical change. The Transitions field has been repeatedly criticised for its failure to explicitly deal with politics and the often political nature of processes of change in sustainable directions (e.g. Smith and Stirling 2007; Smith and Stirling 2010; Kern 2011; Meadowcroft 2011; Lawhon and Murphy 2012). Despite these repeated calls for more attention to the politics of change, and even a contribution by one of the literature's key proponents seeking to extend one of its core conceptual frameworks to attend to politics (Geels 2014), there is only a handful of examples in the Transitions literature where empirical analysis has tried to deal with politics, political economy or power (some examples include: Avelino and Rotmans 2009; Grin 2010; Kern 2011; Normann 2015). In a developing country context, the literature is practically brand new: Baker et al. (2014) analyse the political economy of South Africa's energy transition, and this may be the only example of a peer-reviewed journal paper within the Transitions field to date. Low-income countries also tend to exhibit extreme asymmetries in the distribution of power and of knowledge, both often being the privilege of a centralised political and scientific elite, thus further emphasising the need to attend to politics and power in understanding socio-technical change.

The analysis in this book shows why the ignorance of politics in multiple forms is a fundamental weakness that requires much more attention in future research – and action – in this field. However, rather than seeking to provide the definitive answer to this weakness, we instead use our analysis to emphasise, first, the importance of politics, by showing how it is relevant to explaining the ongoing development of the solar PV market in Kenya. We do this through a close examination of the work done over several decades by key actors who have helped to build a 'socio-technical innovation system' around PV systems, work that is often political as well as technical. In revealing the often political nature of this work, we can take a second step, one that reflects on the implications for both the analysis and governance of action to achieve transformations in sustainable energy access. Important in this regard is this notion of 'key actors', whether they are individuals or organisations. So, to be clear, our aim is to extend the contribution of this book beyond just the articulation and demonstration of a socio-technical innovation system perspective on sustainable energy access. We also want to articulate the importance and implications of attending to the interventions of key actors seeking to realise or build socio-technical innovation systems. However, we are fully cognisant of the fact that our conceptual framework requires further work in this direction. It is here that we begin to make our third step, by outlining what we think a future research agenda could be that would enable us to strengthen the political and governance dimensions of a socio-technical innovation systems framework, both as an analytical and an action-oriented perspective.

Our action-oriented motivation is important because a focus on the work of key actors is of direct relevance to the ambitions of international policy commitments like the UN's Sustainable Energy for All initiative. These commitments imply a need to understand how actors might deliberately intervene to drive 'broad' transformations in sustainable energy access, as opposed to understanding how 'narrow' transitions might evolve (or have evolved) over time (see Stirling 2014, or Stirling 2015a, for a discussion of the distinction). Placing these actors within a systemic perspective on change – one that is assisted by drawing on the Socio-Technical Transitions and Innovation Studies literatures – we refer to such actors as socio-technical innovation system builders. This facilitates close attention to the many international and national actors who seem to play key roles in promoting sustainable energy access in low-income countries, e.g. intergovernmental organisations (IGOs), non-governmental organisations (NGOs), researchers (including us, the authors), private sector actors, technology users, and so on. In myriad ways, they (we) all participate in processes of 'development', change and knowledge co-production and many of them (us) 'move easily between Washington and Addis Ababa, Rome and Bamako' (Keeley and Scoones 2003, p. 163). This raises classic questions of power, legitimacy and distribution: Whose knowledge counts? Who has control over resources? Whose agenda drives change? Who wins? Who loses? Once again, we cannot answer all these questions but we do aim to show that the framework we develop in this book at least provides a way to analyse these questions and to think about how to address the huge asymmetries in power, knowledge and distribution we have noted.

While we have argued that both the Innovation Studies and Socio-Technical Transitions literatures are lacking in their treatment of politics, there is a handful of papers that make reference to relevant ideas through the notion of systemic intermediaries. In general, intermediaries are actors who work across the boundaries between firms or sectors (Howells 2006; Kivimaa 2014) or between supply-side and demand-side actors (Stewart and Hyysalo 2008), providing a wide range of services to clients in primarily bilateral relationships. For example, an intermediary may be employed by a client organisation to conduct a market survey or technology foresight exercise. Systemic intermediaries differ from this general type in that they 'function primarily in networks and systems … and focus on support at a strategic level' (van Lente *et al.* 2011, p. 39). This systemic function, or strategic focus, resonates with our notion of system builders in the sense that they have the potential to transform socio-technical systems (Marvin *et al.* 2011). But similar notions appear under different terms in other, more specifically Socio-Technical Transitions-oriented, work. Here, they are described as actors who play a role in driving cumulative causation (often borrowing from Political Science ideas such as 'advocacy coalitions' and 'policy entrepreneurs') (Kern 2011); as 'technology advocates' who do socio-political work to empower socio-technical niches, including by constructing actor-networks (Smith and Raven 2012); and 'cosmopolitan actors' who do socio-cognitive work to render technologies more widely applicable and lead to their being used beyond sustainable niches (Deuten 2003). However, so far, only Kivimaa (2014) acknowledges the work of systemic intermediaries as political; indeed, she even argues that it is necessarily so. Significant work, then, needs to be done to develop these threads into a comprehensive theory that explicitly deals with the role of such actors and the political nature of their actions. Moreover, this needs to be clearly situated within a systemic and active perspective of how transformations derive from such actions. It is beyond the scope of this book to complete such work. However, our hope is that at the very least the analysis articulated herein goes some way to bringing the significance of the idea of socio-technical innovation system builders to the foreground of analysis, and also articulates their importance to thinking on policy, practice and governance more broadly.

The structure of this book

We develop our 'Socio-Technical Innovation System' conceptual framework and related concept of 'Socio-Technical Innovation System builders' in subsequent chapters. Chapter 2 focuses on relevant insights from Innovation Studies while Chapter 3 focusses on Socio-Technical Transitions. In Chapter 4 and Chapter 5, we move on to demonstrate how a socio-technical innovation system perspective has more explanatory power in understanding one of the most widely hailed examples of low carbon energy technology uptake in Africa, namely, the solar PV market in Kenya. This is achieved by analysing an in-depth reconstruction of the history of solar PV in Kenya, based on a combination of over one hundred hours of recorded interview testimony, stakeholder workshops and extended time spent by the authors in the field, both as researchers and practitioners. In Chapter 6, we focus on two classic examples of alternative solar PV policy approaches that have been implemented in Kenya, which respectively rehearse the Hardware Financing and Socio-Technical Innovation System Building framings we characterise in this book. This analysis demonstrates the stark difference in the transformative impact between the two, where the latter has had significantly greater and more rapid impacts than the former.

The analysis in Chapters 4–6 is focussed specifically on solar PV in Kenya and the findings suggest there is an urgent need to conduct more comparative future research across different scales and types of sustainable energy technologies, and different socio-cultural and political contexts. While this research is yet to be done in a way that builds on the analysis in this book, in Chapter 7 we draw on insights from several other pieces of original empirical analysis in Tanzania, India and China – work with which the authors have been involved. This shows how a systemic perspective resonates across these different contexts and, indeed, across other technologies and different points of enquiry across a range from consumer access to technologies (in Kenya and Tanzania) to industrial activity in relation to sustainable energy technologies (in India and China).

Before giving some final thoughts in this Introduction, we should explain the use of capitalisation for some terms that the reader may already have noticed. Building on Hulme (2009), we adopt the convention of using upper-case letters to denote when we are referring to framings and the associated narratives that support them. So, Private Sector Entrepreneurship in upper case denotes the dominant policy framing and associated narratives that portray private sector entrepreneurship as the key to achieving sustainable energy access, whereas private sector entrepreneurship in lower case simply refers to, or describes, an instance of entrepreneurship observed in the private sector. In keeping with this, we also use upper-case letters to denote broad areas of scholarship, which could be considered themselves to be 'frames' in that they each include and exclude different elements according to their particular perspective. So, the use of Socio-Technical Transitions refers to the corresponding literature and scholarship while socio-technical transitions would refer to specific processes of socio-technical change. We hope this provides some clarity in regard to whether we are speaking at any particular point about framings (or area of scholarship) or whether it is in reference to a specific instance of a particular practice or process. Likewise, we use this convention for our own framings in the hope that it creates transparency in regard to our own position.

Towards pro-poor governance of sustainable energy access

The analysis and theoretical approach in this book, as well as the insights for policy and practice, leave as much to be researched and articulated as they provide any concrete answers to the problem of sustainable energy access. The enormity of the sustainable energy access problem in the Global South is not one that can be solved in one book, nor by any singular prescription for policy and practice. Indeed, even in pursuing the research agenda articulated herein, it is neither our aim nor our desire to achieve any such prescription. Instead, what we hope this book does is to point us in the direction of an approach to the governance of sustainable energy access that embraces the fullest understanding of the notion of sustainability.

Centre stage in this is a concern with social justice and democracy (c.f. Forsyth 2008). If we accept the promise of a systemic perspective on socio-technical change and its relevance for catalysing development that meets the needs of poor countries and poor people, then we also need to consider how we create and nurture supportive institutions and governance structures. For us, this is a call for governance structures and institutions that help to build capabilities, from those at the individual level, such as the kinds of capabilities advocated by Sen (e.g. Sen 1999); to those that are technical and systemic in nature, such as the kinds of capabilities that facilitate technological innovation (e.g. Bell 2012). Simultaneously, these structures and institutions need to foster networks that meaningfully connect different individuals, groups and constituencies at multiple levels (Forsyth 2005; 2007) – connections that enable flourishing and inclusive partnerships in project and programme design, implementation and evaluation (Sovacool and Drupady 2012, p. 295). We offer our concept of Socio-Technical Innovation System Building as a way to realise these goals, an approach that can catalyse change by seeking to understand – across specific but widely varying socio-cultural, political and economic contexts – existing practices, and existing technological strengths and weaknesses, and to build on these through inclusive and reflexive projects, programmes and other interventions.

Of course, even if our approach is accepted in some form, there is still plenty of work to do in order to develop it. We conclude the book, therefore, by offering an agenda for future research, policy and practice in this field. Before doing so – notwithstanding the critiques raised above regarding policy prescription – we do attempt to articulate a concrete policy proposal for how to implement our approach in relation to climate technology interventions; a proposal we have articulated elsewhere (see Ockwell and Byrne 2015), and which we have called Climate Relevant Innovation System Builders (CRIBs). We reiterate the proposal here in order to demonstrate at least one way to operationalise the concepts we develop through the book. It also raises the possibility of thinking about our approach as one that is applicable beyond the specific challenge of sustainable energy access. That is, Socio-Technical Innovation System Building may be relevant to a whole range of sustainability challenges, not just those in the realms of energy access or climate change. Whether our approach has wider applicability or not, it is our hope that the suggestion of a step change in the way we understand sustainable energy access – and the processes of change that will accompany (indeed, are accompanying) the many interventions aimed at driving lower carbon pathways of development – will form the basis for governance processes more directly focussed on, and preoccupied by, empowering and enabling plural voices – in particular, the voices of the poor and marginalised countries and the people therein.

Notes

1   In this book, we are specifically interested in low-income countries, with a particular focus on Sub-Saharan Africa. For ease of reading, the term 'Africa' is sometimes used. In no way is this intended to portray 'Africa' – or indeed 'Sub-Saharan Africa' – as a homogeneous entity. The critical importance of the myriad cultural, historical, political, etc., heterogeneities that characterise different national and sub-national contexts across Africa is central to our perspective. The fact of these heterogeneities forms a key part of our analytical foci and a core component of our argument for the need for a new theoretical framework, based on detailed, comparative analysis within specific contexts in Africa.
2   See www.teslamotors.com/en_GB/presskit
3   For a more detailed treatment of the issue of low carbon technology transfer, see the various contributions in Ockwell and Mallett (2012).
4   Elsewhere we have explored the gender implications of the emerging focus on entrepreneurship for delivering low carbon technological change in terms of reinforcing existing gendered power relations, see Marshall *et al.* (in press).

2

INNOVATION SYSTEMS FOR TECHNOLOGICAL CHANGE AND ECONOMIC DEVELOPMENT

Introduction

In Chapter 1 we introduced the idea of a theoretical framework that combines relevant insights from the Innovation Studies and Socio-Technical Transitions literatures used within the broader heuristic of the Pathways Approach (Leach *et al.* 2010a). The core aim of this endeavour is to develop a conceptual framework that can underpin research and policy thinking that is better able to support the transformative goals of initiatives like the UN's SE4All and its commitment to achieving universal sustainable energy access by 2030. In this chapter, and in Chapter 3, we take some time to look in more depth at aspects of these two literatures and draw out the threads that support our conceptual framework, with its emphasis on building Socio-Technical Innovation Systems as a means to effect transformative change. The current chapter focusses on relevant insights from Innovation Studies, looking, in particular, at this literature's engagement with systemic perspectives on technological change and innovation. We focus mostly on the National Innovation System (NIS) literature as a lens through which to understand the role of technological change in relation to economic development, including comparisons between different countries' 'levels of development' and the ways in which processes of 'catching up' have been achieved, particularly in the Asian Tiger and rapidly emerging economies.

Chapter 3 then goes into more depth on relevant insights from the Socio-Technical Transitions literature. This enables us to go beyond the firm-centred focus of the NIS literature (and the Innovation Studies literature in general) and engage directly with users of sustainable energy technologies in developing countries, the social practices that these technologies are intended to facilitate, and the co-evolutionary dynamics between social practice and innovation and technological change. This extension also allows us to directly consider the socio-technical regimes with which sustainable energy technologies must compete.

At the end of Chapter 3 we also note what each literature has to say about the role of key actors who deliberately intervene to facilitate change. This, as emphasised in Chapter 1, provides us with a way to analyse the potential role that interventions through policy and practice might play in driving the kinds of transformations implied by SE4All, as well as those of multiple other international policy initiatives. In the case of the Innovation Studies literature, discussed in this chapter, these key actors are conceptualised as various kinds of 'intermediaries' (van Lente *et al.* 2003; Howells 2006). In the Socio-Technical Transitions literature, discussed in Chapter 3, these actors go by different names and are considered to play different kinds of roles. For example, we see them described as 'technology advocates' or 'cosmopolitan actors', among other terms. As we show in the extensive analysis of the Kenyan PV case in Chapter 4 and Chapter 5, these actors can play many different kinds of roles. What emerges from our case is that we do not yet have satisfactory theory to be able to conceptualise the wide range of activities involved, how these activities contribute to Socio-Technical Innovation System Building and, crucially, what it all means for governance. We cannot, in the space of this book, answer all these questions or resolve the many issues that arise. Instead, in the final chapter, Chapter 7, we can point to an agenda for research – and policy and practice – that begins by acknowledging the potential for Socio-Technical Innovation System Building (and builders) and then articulate what we believe are some of the key directions for further work.

An innovation systems perspective: fertile gardens for sustainable transformations

Before going into further details on several key aspects of the Innovation Studies literature (in the context of sustainable energy access and low carbon development more broadly), it is useful first to briefly define what we mean by 'innovation systems' and to outline their relevance in the context of this book. The concept of innovation systems emerged from a body of work in the 1980s and 1990s that sought to define an alternative perspective to neo-classical theories of economic growth (Nelson and Winter 1982; Freeman 1987; Lundvall 1988; Patel and Pavitt 1995; Freeman 1997). Essentially, neo-classical economics was challenged on the basis of its failure to account for the role that innovation and technological change played in driving economic growth, a role that proponents claimed the NIS perspective could better understand (for a historical overview of the NIS literature and its treatment of development, governance and politics, see Watkins *et al.* 2015).

An NIS perspective emphasises the context within which processes of technology development, transfer and adoption occur and, through this, offers a more systemic understanding of technological change and economic development than neo-classical economics accounts. Essentially, an NIS refers to the (primarily national) network of actors, and the strength and nature of the relationships between them, from which both innovation and technological change emerge. It is important to note that the NIS perspective considers 'formal' actors rather than individuals and so includes those players such as firms, universities, research institutes, government departments, non-governmental organisations (NGOs), and so on. It is also important to note that innovation and technological change as referred to here would include processes of technology development and technology transfer, two terms that have been enshrined in international climate change legislation under the United Nations Framework Convention on Climate Change (UNFCCC) and are thus used widely in international policy discourse.

The NIS perspective was based initially on empirical studies in OECD country contexts (OECD 1997), but this was followed soon after by work on the Asian Tiger economies and, more recently, on rapidly emerging economies such as China and India. As a result, the NIS perspective now has a large and diverse collection of evidence that supports its emphasis on innovation systems as the means through which to facilitate both innovation and technological change, and hence economic development. The empirical work in developing countries, as Watkins *et al.* (2015, p. 1410) put it:

> offered some alternatives to how successful innovation systems might be con-structed for purposes of both catching up and for sustained economic growth. For developing countries, this provided both a potential policy roadmap for development, while at the same time laying bare the stark institutional disparities between the developed north and much of the developing south.

Different streams of the NIS literature have subsequently placed particular emphasis on innovation systems as an explanatory approach at the level of specific technologies (the Technology Innovation Systems literature, e.g. Hekkert *et al.* 2007), sectors (the Sectoral Innovation Systems literature, e.g. Malerba and Mani 2009), and regions (the Regional Innovation Systems literature, e.g. Cooke *et al.* 1997).

While interesting in and of themselves, these various streams of the literature on innovation systems are not of core interest to us in this book. What is of relevance is the idea of foregrounding a systemic perspective on economic development in a way that takes seriously the role of innovation and technological change. This emphasis on innovation and technological change speaks directly to current and emerging visions of sustainable transformations (in both the academic literature and international policy discourse) and the extent to which they imply a key role for the develop-ment and adoption of sustainability-enhancing technologies. This is clearly the case in relation to this book's central concern with sustainable energy access, but equally applicable across the suite of areas with which international commitments, such as the new Sustainable Development Goals, intersect (e.g. pharmaceuticals, agriculture, information and communication technologies (ICTs), etc.).

One way to grasp the concept of innovation systems and the role they play is to think of the analogy of a garden. Essentially, innovation systems can be thought of as fertile gardens within which technological change and economic development can prosper. Building on this analogy, we can also introduce several other key insights from the Innovation Systems literature and Innovation Studies more broadly. First, the literature privileges a focus on building technological capabilities in developing countries as the means through which more rapid technological change and development will occur. This is emphasised over and above a narrow focus on individual pieces of technological hardware; the latter being the main fixation of international policy foci on technology transfer (see Chapter 1). Technological capabilities, as described by the Innovation Studies literature, might be thought of as the soil in the innovation system garden. Without attempts to nurture the soil's fertility, scattering seeds (bits of technology hardware) is unlikely to lead to a flourishing garden (technological change and economic development). Moreover, commercial gardening contractors (technology investors) are unlikely to invest effort in sowing seeds in unfertile gardens in the first place – telling us something fundamental about where international policy efforts might focus, if indeed their aim is to contribute to transforming local contexts within which sustainable energy (or other) technologies might make a positive, pro-poor difference.

In the subsequent sections of this chapter we deal with each of these component parts of the Innovation Studies literature. In order to avoid any misconception or misreading of this book, however, it is first necessary to explain exactly what we mean by 'innovation'. This is to pre-empt any misconceptions that equate innovation with 'invention' or, indeed, tend towards an understanding of innovation as connoting radical innovation, as opposed to the more common (and essential) incremental and adaptive nature of innovation and technological change that underpins economic growth and development.

What is innovation?

For many people, including many professionals working in research, journalism and policy, the word *innovation* is understood and used in a way that is synonymous with *invention*. For example, the African Union's heads of state and governments recently adopted the Science, Technology and Innovation Strategy for Africa 2024 (STISA-2024) – a decadal vision aiming to place science, technology and innovation (STI) at the 'epicentre of Africa's social and economic development' (African Union 2014, p. 8). The strategy met with various critiques that were summarised in a review in the leading global science journal *Nature*. This included critiques such as the strategy's having a top-heavy administrative structure; that it is lacking firm commitments from governments to funding or training; its having aims that are beyond the limits of the continent's existing financial resources; failing to provide sufficient detail on how the vision will be achieved; and creating new institutional structures as opposed to building on existing ones (Nordling 2014). However, in both *Nature*'s coverage of reactions to STISA-2024 and in the strategy document itself, innovation is erroneously treated as synonymous with invention or, at best, synonymous with early-stage research and development (R&D). So, in *Nature*'s coverage, we see mention of the failure of African nations to meet their target of at least 1 per cent gross domestic product (GDP) spent on R&D, and graphs comparing low levels of R&D spending in African countries with the much higher levels of R&D spending in the USA and Europe. Certainly, R&D forms part of the STI fabric of an economy but, as we argue below, it is only one small part. This is increasingly recognised by various international bodies interested in measuring innovation, even if efforts to develop indicators that satisfactorily capture the wide variety of innovation activities that contribute to economic development are far from complete (WIPO 2011; Cornell University *et al.* 2015).

One of the major difficulties that any of the organisations cited above has encountered when attempting to measure the wide variety of innovation activities is that these activities contribute to outcomes that are not just about new pieces of technological hardware. As Stirling (2015b, p. 1) asserts, innovation is not just about technological invention; 'It involves change of many kinds: cultural, organisational and behavioural as well as technological.' This definition points to the fact that innovation is something that goes well beyond technological hardware, and well beyond the activities of firms, industries, universities and research centres. We will return to this broader relevance of the term 'innovation' further below and in Chapter 3. For now, let us first deal with innovation in the context of technological change in relation to firms and industries, for, even here, a proper understanding of the meaning of 'innovation' goes well beyond the common assumption of inventing technologies that are new to the world.

Fagerberg (2005, p. 4), for example, states that *invention* is considered to be the first occurrence of an idea (e.g. how to harness certain technical principles to make a touchscreen interface), while *innovation* is considered to be the first implementation of that idea in practice (e.g. the incorporation of a touchscreen into a new mobile phone released on the market). Furthermore, not all innovations are radically new technologies based on scientific R&D. As the OECD's *Oslo Manual* asserts (OECD 2005, pp. 46–47), it is also innovative when a firm is the first to introduce a new (or improved) piece of hardware, e.g. a product or piece of production equipment, or a new (or improved) technique, e.g. a production process or marketing strategy. Likewise, even if other firms have already introduced new hardware or techniques, it remains innovative to a firm when it adopts these itself for the first time. Indeed, as Arnold and Bell (2001, p. 288) note when discussing product innovation in firms in OECD countries, 'considerable efforts are devoted to monitoring competitors' products and reverse engineering – both as a source of ideas and in order to benchmark the company's own processes'. Moreover, Arnold and Bell (2001) report that these kinds of efforts constitute the dominant form of innovation activity, contrasted with the 'vanishingly small' contribution of publicly-funded R&D. This does not provide an argument against public sector research; rather, it provides a perspective on the direct contribution of such research – particularly so-called basic science – to innovations in economic activity.

Where publicly-funded scientific R&D does appear to be more useful to an economy, according to Arnold and Bell (2001), is in the provision of highly-trained researchers who can then work in private firms to further innovative activities. One of the reasons for this benefit is that such researchers are able to understand the knowledge created by scientific R&D and so raise the chances of applying that knowledge in the firms' innovative activities. The ability of a firm to understand and make use of scientific R&D, and of the stock of knowledge more generally, is referred to as 'absorptive capacity'[1] (Cohen and Levinthal 1990, p. 128). This ability to use scientific knowledge is likely to be more important in those areas of economic activity that are, broadly speaking, at the frontiers of technology, e.g. bio- and nanotechnology, among others. Indeed, in such frontier areas, where the risks to investment are high, publicly-funded research ('basic' and 'applied') can be critical to the initial development and future success of those technologies, as well as the firms that work in their respective sectors. As Mazzucato (2013, p. 13) argues:

> From the development of aviation, nuclear energy, computers, the Internet, biotechnology, and today's developments in green technology, it is, and has been, the State – not the private sector – that has kick-started and developed the engine of growth, because of its willingness to take risks in areas where the private sector has been too risk averse.

Once innovations are introduced or adopted by firms, there often follows a continuous process of improvement (e.g. in efficiency) or adaptation (e.g. to meet the regulatory requirements of another country) in which each implemented change is also considered an innovation. Such *incremental* changes can add up to significant improvements over time, as Barnett (1990, p. 543) observes:

> much of the increase in productivity in industrialized countries is achieved through the aggregation of myriads of minor changes to existing production processes (rather than from individual massive jumps in productivity through investment in new vintages of technology).

We return to the importance of this kind of incremental innovation further below.

Adaptive innovations are also often required in order to ensure an existing innovation better 'fits' the context into which it is introduced – a new country, industry, firm, farm, household, etc. – such that it is more likely to be adopted or that it performs better in that new context. For example, many adaptive innovations have been made to mobile phone handsets being sold to poor consumers in Kenya. Foster and Heeks (2013, p. 343) describe the innovation responses of Chinese mobile handset firms to suggestions from Kenyan intermediaries working close to low-income consumers for modifications to handsets:

> [Innovations included] dual sim card phones (allowing users to choose the lower-cost network to phone particular contacts), translation of the phone interface into Swahili, and addition of a single-button-enabled new interface for the popular M-Pesa mobile money service.

It should be clear from this brief discussion that there is a wide spectrum of outcomes that can be described as innovations. The latter example also brings us back to the implication of Stirling's (2015b) broader definition above, i.e. that innovation is about change beyond just technological innovation. As the Chinese mobile handset example illustrates, technological innovation often (sometimes deliberately, sometimes more organically) co-evolves with the social practices that technologies are intended to facilitate, or sometimes 'accidentally' enable. Understanding such co-evolution is central to the literature on Socio-Technical Transitions dealt with in Chapter 3.

Innovation as process and outcome

Sticking, for now, with a focus on firms and industry, as the majority of the mainstream Innovation Studies literature does, we see that the term 'innovation' is also often used to refer to the *process* of innovating – as opposed to just the outcome – and it is understanding this process that is of fundamental concern to much of the Innovation Studies literature (Fagerberg 2005, p. 9). In many cases this focus on the processes of innovation has an explicit or implicit goal of better understanding how innovation processes might be managed. This could be in order to inform how firms might improve efficiency or profitability, or it could be to inform public policy design geared towards supporting the development and competitiveness of domestic industries. Of particular interest to us in this book is the concern in the Innovation Studies literature that has focussed on analysing and explaining the evolution of firms, industries and whole economies in the developing world (e.g. the Asian Tiger economies and more recent emphasis on rapidly emerging economies such as China and India).

The outcomes of innovation processes are inherently uncertain and therefore unpredictable, except in broad terms. However, it is clear that innovation outcomes do not occur simply by chance and that decades of research into innovation processes have given rise to many useful insights. At a general level, one of the most important of these insights has been the recognition of the systemic nature of innovation, which has driven the development of increasingly sophisticated 'models' of innovation. In the literature, the so-called 'linear model' has long been discredited. This model assumes innovation begins with scientific research, which leads to innovations of various kinds via a sequence of steps through engineering, manufacturing and marketing. During the 1960s, this 'science push' understanding of innovation gave way to a 'market pull' model, which emphasises the importance of needs expressed in the marketplace as motivators of innovation, although it, too, is a linear representation of the innovation process (Rothwell 1994). Based on the findings of many detailed empirical studies of innovation in practice, which revealed the inadequacies of both these linear conceptualisations of innovation, Kline and Rosenberg (1986) introduced a coupling model, which forms the basis of the innovation chain we now see widely cited. The coupling – or interactive or chain-linked – model is shown in Figure 2.1 with the addition of the existing stock of knowledge available to firms, which Arnold and Bell (2001, p. 287) argue continues to constitute the vast majority of knowledge used during innovation activities.

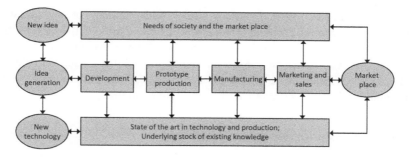

FIGURE 2.1 An interactive model of innovation
Source: Based on Kline and Rosenberg (1986, p. 290), Arnold and Bell (2001, p. 287) and Conway and Steward (2009, p. 68).

Although this interactive model emerged from empirical analysis of firms, its arrangement of different kinds of functional activities in innovation – linked to the marketplace and to the available stock of knowledge – suggests not all activities need to be performed within any one firm. Instead, the empirical studies revealed the reality of multiple feedbacks between these functional activities and the available stock of knowledge and society where the feedbacks are represented by the two-way arrows in the model. Moreover, innovations could emerge from any one of the activities – say, marketing and sales – or any combination of some or all of the activities. From this perspective, the nature and quality of the feedbacks and links between activities become as important as the activities themselves, regardless of whether the activities are located in a single firm or in many specialised firms. For example, the manufacture of a product could be excellent in terms of quality, and it might be skilfully marketed, but if few in the marketplace are interested in that product, then it is likely to be a failure. Conversely, a poorly-made product, even if in principle it answers a widespread need in society and is well marketed, is likely in time to fail. In the first case, we might surmise that poor understanding of particular needs in society (because of weak feedbacks and links) led to poor design and manufacturing choices. In the second case, despite strong feedbacks and links that enabled a clear understanding of needs, the manufacturing ability was too weak to answer those needs satisfactorily.

So, we can see that the concept of innovation encompasses both the *process* of innovating and the *outcome* of this process. We also see that the outcomes – innovations – are not always radically new technologies or techniques; indeed, the majority of innovations are incremental or adaptive, emerging from interactive processes anywhere in the innovation 'chain', not just from – or even necessarily involving – basic R&D. This more sophisticated model of innovation activities, together with an appreciation of the wide spectrum of innovation outcomes, provides some of the essential features that need to be understood in order to make sense of what we now call 'innovation systems'. But there are other foundational ideas from the Innovation Studies literature that we need to highlight before we can draw the various insights together into a broad understanding of an innovation system. Most importantly, we need to explore the relationship between knowledge and technology (and, more generally, innovation), which has important implications for understanding the means through which policy might intervene to encourage the transfer, development and adoption of sustainable energy – or other sustainability-relevant – technologies in developing countries.

Technology as knowledge not hardware

A critical insight from the Innovation Studies literature is that technology is not simply hardware. Embedded in the hardware is a reflection of the knowledge required to create it; and the knowledge and skills (sometimes referred to as the software) which are needed to adopt, use and adapt that hardware (Bell and Pavitt 1993; Ockwell *et al.* 2010b). An essential characteristic of this 'software' is tacit knowledge – a fundamental aspect of knowledge and skills that is difficult or impossible to articulate or codify, but an aspect that can be cultivated through practice and experience (Polanyi 1966) (see below for an elaboration on codified versus tacit knowledge). Taking these insights together with those discussed in the previous section, we can begin to craft a more holistic understanding of how technologies are created, adopted, used and adapted. And, from this holistic perspective, we can see that understanding technology (and, more broadly, innovation) in terms of knowledge has profound implications for how technology development and innovation can be more successfully – and sustainably – encouraged than the currently dominant approaches of hardware financing.

The chain-linked model discussed above (see Figure 2.1) derives from the recognition of the importance of inter-linkages between firms, and their inter-linkages with the social context. This already suggests interdependent relationships between firms, innovations and society. Innovations succeed, in part, because they respond to well-understood 'needs' (or demands) in the relevant social context and so we can see that the behaviour of firms (and the nature of innovations) are to some extent dependent on that social context (a point developed in more depth in Chapter 3 in relation to Socio-Technical Transitions perspectives). But, as innovations become more widely adopted, they enable new choices, behaviours, and so on, in society (including among innovating firms). So the context is also dependent to some extent on the behaviour of firms and the innovations they produce. And, of course, all this happens within an environment of institutions constituted by policies, laws, regulations and norms. Furthermore, these institutions evolve as responses to, or intentions to influence, developments in technology and other innovations. In other words, technologies (innovations) can be considered part of a system-like whole, and a key *dynamic* element of that system is evolving knowledge (including scientific, technical, social, cultural, political, etc. forms of knowledge). A particular technology or innovation can be understood as a specific distilled combination of these different knowledge domains, including what can be codified and what is inherently tacit. Within the traditional Innovation Studies literature, many of these ideas have been thoroughly analysed and there is, in particular, a wealth of evidence concerned with the relationships between evolving knowledge, firms' behaviour and the formal institutional environment (e.g. see Katz 1987; Kim *et al.* 1989; Bell 1990; Freeman 1992; Lundvall 1992; Bell and Pavitt 1993; Hobday 1995a; 1995b; Bell 1997; Radošević 1999; Ockwell *et al.* 2008; Bell 2009; Watson *et al.* 2015).

One way to understand the significance of these ideas in relation to the transfer of technologies to developing country firms (be they sustainability-relevant technologies or otherwise) is depicted in Figure 2.2, especially in regard to innovation systems and the ways in which the knowledge and skills required for self-directed development can be accumulated. Adapted from Watson *et al.* (2015, based on Bell 1990), Figure 2.2 shows three types of technology transfer flows (A, B and C) into a local context. Flow 'A' includes hardware, as well as the engineering and managerial services that are required to implement such transfer projects. Flows of type 'B' consist of information about production equipment – operating procedures, routines, etc. – and training in how to operate and maintain such hardware. Bell (1990, p. 77) describes these flows as 'paper-embodied technology' and 'people-embodied knowledge and expertise'. Both flows 'A' and 'B' add to or improve the production (and, we might add, consumption) capacity of a firm or economy, but do little to develop the skills needed to generate new technology. The hardware financing policy mechanisms characterised in Chapter 1, like the CDM, tend to result in these kinds of flows: for example, the CDM might facilitate the installation of wind farm projects that increase the capacity for electricity production; the diffusion of compact fluorescent lamps enabled by the CDM's Programme of Activities stream enables more efficient consumption of electricity for lighting (e.g. Byrne 2013a). Flows of type 'C', however, are those that help to create the capability to generate new technology. In other words, they help to build 'innovation capabilities' (see Bell 2009). These flows have not occurred through the CDM without the active leveraging of project opportunities through the implementation

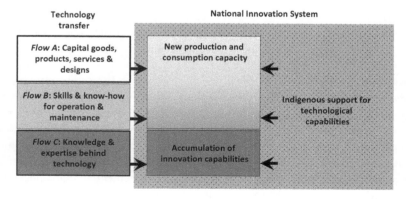

FIGURE 2.2 Technology transfer and indigenous innovation
Source: Adapted from Watson et al. (2015).

of additional national policy interventions, such as those used in China (Stua 2013; Watson *et al.* 2015; and see the case study on China in Chapter 7).

Within the context of a concern with sustainable energy access and broader concerns with, for example, low carbon development or climate technology transfer, this idea of technology flows building local capabilities to generate broader technological change is of central importance. In the latter case, it is important for building capabilities to generate technological changes that facilitate more rapid engagement with low carbon technologies by firms in developing countries. In other words, efforts need to focus on building technological capabilities in developing countries, moving from basic production and consumption capabilities (the ability to use existing technologies) towards innovation capabilities (the ability to build on access to existing technologies to manage processes of innovation and change). The existing technological capabilities in the local context could also be considered absorptive capacity, broadening Cohen and Levinthal's (1990) original definition beyond the capacity of a single firm. Indeed, this broader concept has been used to demonstrate the impact of individual firms' absorptive capacities on the ability of clusters of firms to adopt and adapt new technologies (Giuliani and Bell 2005), and to explain the ability of countries to achieve technological learning through the CDM (Doranova 2010).

Figure 2.2 does not explicitly show the importance of the institutional environment, although the Innovation Studies literature does so, especially with regard to formal national and international institutions. Used carefully, these can help to enhance existing industrial activity – e.g. to raise the level of capabilities to increase competitiveness – but they are also important for fostering new industrial activity that would otherwise not be pursued (e.g. see Cimoli *et al.* 2009). In the case of sustainable energy technologies – and a concern with broader processes of sustainability-related technological change – this latter point is particularly relevant (Ockwell *et al.* 2010b). Many existing sustainable energy alternatives cannot yet compete with unsustainable, mainstream (usually fossil fuel-based) technology options and so market demand for many sustainable energy technologies tends to be weak or marginal. But it is likely that we will need a range of such technologies, and the need is becoming increasingly urgent. In principle, appropriate policies could foster their competitiveness, and the local capabilities (and innovation systems) that can sustain and develop them.

It should be clear from this discussion that achieving large-scale technological change is usually an inherently long-term endeavour, involving multiple actors engaged in many interdependent processes. The literature that analyses these changes – particularly the literature on the so-called 'catching-up' countries – demonstrates convincingly that making the transition from a least-developed country to even a middle-income developing country status requires targeted strategic policy interventions implemented over decades (e.g. Fagerberg and Godinho 2005). While there are many context-specificities (see more detailed discussion further below) that mean each country has implemented particular policies might not be suitable elsewhere, the general strategy has been to ensure the development of technological and innovation capabilities and to build innovation systems (Cimoli *et al.* 2009). As we have highlighted, central to this endeavour is the exploitation of knowledge flows. This has been a huge challenge for those countries that have – at least to some extent – 'caught up'. The challenge that a transformation in access to sustainable energy (or other sustainability-related) technologies faces is the need to build relevant innovation systems and to build them quickly – constraints that did not trouble the previous 'latecomers' (Byrne *et al.* 2014a). Nevertheless, there is evidence in the literature that suggests many practical policy interventions to achieve such development are available, although we should also expect there to be a continuous need for experimentation and learning (Chaminade *et al.* 2009).

The incremental nature of most innovation and technological change

The discussion further above alluded to the importance of the incremental nature of the majority of innovation. Here we expand on this issue with reference to some empirical examples in order to clearly illustrate its policy relevance. In line with the popular conception of innovation as science-based radical invention, there is a tendency to assume that the promotion of innovative capabilities will be readily achieved through the strengthening of science-based R&D. However, experiences reported in the literature suggest that this strategy is unlikely to work unless more foundational capabilities are built first. For example, Bell (1997) argues, with reference to a number of studies – from as early as the 1970s – focussed on a range of countries in Latin America and East and South-East Asia, that knowledge acquisition and capability-building are better achieved through incremental improvements that begin with simple engineering and managerial competences. Citing the work of Katz and colleagues (Katz 1987), and research conducted by Hobday (1995a; 1995b) and Kim *et al.* (1989), Bell discusses two different sets of approaches to establishing inter-firm and organisation linkages; one based on centralised R&D services and the other more sequenced along a 'simple' to 'complex' trajectory. The Latin American countries used a centralised approach in which specialised R&D organisations were expected to conduct research on behalf of private firms. This was unsuccessful, largely because the R&D organisations and private firms were unable to communicate with each other. That is, private firms, without sufficient existing technological capabilities, were unable to articulate their technical needs in a form that the R&D organisations could use to focus their research. For their part, the R&D organisations were not necessarily interested in the technologies that the firms were using, especially as they were attracted to the prestige of contributing knowledge to the international scientific frontier. So, while there were elements of an innovation system in place – firms seeking to service local market demands and research organisations with a remit to improve technologies for those firms – the 'system' was poorly articulated and the firms lacked sufficient absorptive capacity.

By contrast, the strategies used in the Asian countries proved to be more successful. These countries tended to use a sequencing approach in which firms were encouraged (and supported through government policy) to first develop their basic engineering and managerial competences. Complemented by other measures, such as huge investments over many years in training thousands of engineers (Freeman 2002), fostering links between firms and targeting a narrow range of strategic industrial sectors (Sauter and Watson 2008), the Asian countries encouraged their innovation systems to evolve from simpler imitative innovation to more complex creative innovation. For example, by the 1990s, Korea had developed a world-leading steel industry, having begun with no capabilities in steel manufacturing in the 1960s. Achieving this included the use of protectionist measures to allow indigenous firms to strengthen their capabilities before having to face the full force of international competition, although the state itself applied significant pressure to ensure that indigenous firms did indeed achieve improvements in their capabilities (for more on these strategies and measures, see discussions in Chang 2002; Reinert 2007; Khan and Blankenburg 2009; Schmitz et al. 2013). China is one of the most iconic current examples of a successful catching-up country. Once again, as summarised in the case study on China in Chapter 7, this success has been achieved to a large extent by concentrating on building technological capabilities in strategic industries and building a well-functioning innovation system (see also Watson et al. 2015).

Why intellectual property rights are a distraction

As emphasised above, knowledge flows are a core component of technology transfer and are essential to building technological capabilities. A core function of innovation systems is therefore to create the enabling conditions to nurture such flows, between firms and other key actors within a country and with relevant actors internationally. However, understanding the nature of these flows, and the relevance of different types of knowledge, is critical to avoiding a fixation on particular policy foci that are unlikely to yield significant benefits in terms of building technological capabilities within developing countries. In particular, differentiating between tacit and codified knowledge, and the greater significance of the former to building technological capabilities, is essential.

Codified knowledge, as the name suggests, relates to knowledge that is articulated in some way. This can include intellectual property rights (IPRs: legal rights over ideas, including copyrights, trademarks and patents), but it also often includes a range of other proprietary knowledge such as trade secrets that have not necessarily been formally or legally protected. The latter overlaps with the second type of knowledge – namely, tacit knowledge. Tacit knowledge refers to human-embodied knowledge acquired through experience of doing things, and would extend to more institutionally embodied knowledge where firms and other organisations develop capabilities around, for example, management systems and approaches that are passed on through generations of employees. Thus, codified knowledge might relate to engineering and manufacturing processes (e.g. for manufacturing advanced wind turbine blades or drought-tolerant crops) whereas tacit knowledge would relate to the applied engineering, systems integration or plant breeding and modification *skills* necessary to effectively work with a new engineering, manufacturing or biotech process.

Even when providing simple definitions of tacit and codified knowledge, it is immediately obvious that relevant tacit knowledge is a prerequisite for codified knowledge to have any use or relevance to a firm or industry. Imagine, for example, giving a local car mechanic access to the patents for core components of a new design of a Formula One racing car engine. It is highly unlikely that the mechanic would be able to successfully build the car without access to the engineering, design and mechanical experience of specialised Formula One development and manufacturing teams (the latter being tacit and neither codified nor legally protected in any way). It is often the processes of tacit knowledge acquisition that more accurately characterise learning by firms and other actors as innovation systems increase in levels of sophistication and scale and lead towards processes of innovation and change. Lundvall (1992) characterises these tacit learning processes into three categories: (1) learning by doing (e.g. by engaging in production processes); (2) learning by using (e.g. making adjustments to get new technologies to fit specific tasks); and (3) learning by interaction as actors interact and work with other actors across innovation systems (see Foster and Heeks 2013). None of these categories of learning and capability development relies exclusively on access to codified knowledge; rather, they represent far more fundamental processes of tacit knowledge sharing and deeper learning.

However, the centrality of tacit knowledge, or experience of working with the technology and processes in question (or related technologies), has tended to be overlooked in international discussions on climate technology development and transfer in favour of a fixation on codified knowledge, and IPRs in particular. As a result, IPRs have become a politically contentious issue within negotiations under the UNFCCC. The debate tends to be framed around two opposing perspectives (Ockwell *et al.* 2010a). On the one hand, some (often developing country) parties and observers claim that developing countries must have free access to IPRs for climate technologies. On the other hand, other (often industrialised country) parties and observers claim that the key barrier to climate technology transfer to developing countries is their weak IPR protection regimes which, it is argued, deter international technology-leading firms from deploying new climate technologies in those countries. The international prevalence of this debate on the role that IPRs do or do not play, and the extent to which international policy should therefore focus on them, continue to dominate many discussions on technology and innovation in relation to international sustainability commitments, probably none more so than under the UNFCCC. For this reason we will now take some time to unpack this issue in detail.

In a close analysis of the background to these two conflicting perspectives on IPRs in relation to climate technologies, and based on a comprehensive review of available empirical evidence that covered a suite of clean energy and energy-efficient end-use technologies, Ockwell *et al.* (2010a) demonstrate three key things. First, the two perspectives can best be understood as having emerged from alternative motivations for industrialised and developing countries to become party to the UNFCCC. For industrialised countries, the key driver was a desire to avoid future economic costs of climate change. This had to include climate (particularly low carbon) technology transfer to developing countries to mitigate future emissions from economic development in these countries (simultaneously providing opportunities for offsets against industrialised countries' own emissions). For developing countries, a core incentive was the promise of access to new technologies – technology ownership being directly correlated with economic wealth and still largely weighted towards the Global North. Without understanding these background motivations, it is difficult to move beyond the political impasse concerning IPRs.

The second insight is that there is empirical evidence to support both sides of the IPR debate. On the one hand, in none of the cases analysed did developing country firms lack access to the technologies in question, and none reported IPRs as having constituted a barrier to technology access. Anecdotal evidence tended to reinforce the centrality of tacit knowledge (as opposed to IPRs) as the key barrier, e.g. with Indian light-emitting diode (LED) manufacturers at the time reporting lack of experience of white spectrum LED manufacturing processes being the key barrier to their entering this new market (not access to patents). On the other hand, several of the firms reported that, were they to attempt to reach the technological frontier in their sectors (e.g. thin film solar PV as opposed to conventional PV panels), IPRs would become a more significant consideration.

The finding that empirical evidence supports both sides of the IPR debate is confirmed in more recent analysis by Abdel Latif (2012). Based on a collaborative project by the United Nations Environment Programme (UNEP), the European Patent Office (EPO) and the International Centre for Trade and Sustainable Development (ICTSD), which included analysis of patent databases and an extensive survey of developing and industrialised country firms involved in clean energy technologies, Abdel Latif (2012) reports that IPR protection was not found to be a significant barrier to technology transfer. Firms interested in licensing to developing countries were seen to be more concerned about other attributes – including favourable markets, the investment climate, human capital and scientific infrastructure – and many firms were willing to offer flexible terms to developing countries that had constrained capacities. This links directly to the final insight from Ockwell et al.'s (2010a) analysis, and the key point in the context of understanding the role and importance of a focus on building innovation systems. That is, focussing on building technological capabilities, which necessitates a focus on innovation system building, is the best way to achieve more sustained development and transfer of climate technologies to developing countries. Thus, the background motivations of both industrialised and developing countries could be better served by concentrating on innovation system building in poorer countries; climate technology access and economic development can be achieved, mitigating future emissions while simultaneously underpinning long-term climate-compatible development.

This brings us to the crux of the matter in relation to the focus of this chapter. Nurturing innovation systems around sustainable energy technologies in developing countries is essential to achieve more sustained processes of self-defined, context-sensitive, sustainable energy technology development, transfer and adoption. These innovation systems are necessary in order to build the technological capabilities of various actors within any given country, taking into account the different context-specific needs of different countries, firms and communities (see discussion below of context-specificities). Knowledge flows are critical to building these capabilities. However, the qualitative type of knowledge that is relevant and likely to have most impact on the capabilities of countries to which knowledge is flowing will vary according to the wide range of context-specific considerations discussed below. In some cases, knowledge will be codified, and in some cases this codified knowledge will be in the form of IPRs (as opposed, for example, to trade secrets). But IPRs are only a small part of a much bigger picture. Access to IPRs does not guarantee developing countries access to sustainable energy (or other) technologies. The cultivation of tacit knowledge is often more important. In many cases, tacit knowledge and knowledge acquired through working with technologies, albeit often under licence and protected by patents, have played a far more significant role than access to IPRs per se. IPRs are only likely to be prohibitive once developing country firms reach the technological frontier in relation to any given technology. As a focus for policy mechanisms aiming to build innovation systems around sustainable energy technologies, IPRs are thus necessary sometimes, but never sufficient.

Context-specificities in technology needs and knowledge flows

A final consideration that policy interventions regarding sustainable energy technologies must negotiate is that of a needs-based approach to policy so as to respond to the context-specificities that define the appropriateness of technology options and related knowledge flows in any given situation (Ockwell and Mallett 2012; Sokona *et al.* 2012; Tawney *et al.* 2015). Hulme (2008; 2009) demonstrates the importance of context-specificities through his discussion of how 'climate change' has been dominated by a particular construction of the issue – a construction that ignores the multiple spatially and culturally contingent understandings and meanings of 'climate'. The implication is that this single-view dominance potentially undermines constructive ways forward for society to both interpret and decide how to respond to a changing climate. This has further implications when considering knowledge transfer between different contexts: it implies contingencies both in terms of what kind of knowledge and related technology might be appropriate, and in terms of the type of policy intervention that will be effective in brokering knowledge transfer and technological capability building. These issues are unpacked below.

In operationalising an innovation systems perspective, one immediately obvious level at which context-specificities matter is in the differences between NISs across developing countries. The decades of the post-World War II era have seen a divergence in the fortunes of developing countries. Some have achieved impressive industrialisation and economic growth, while others have stagnated. Of course, there are many reasons to explain this divergence – including war versus peace, environmental crises versus relative stability, disastrous political leadership versus more benign states, different colonial legacies, rich versus poor resource endowments – but their success or otherwise in building functioning innovation systems is also an important factor (Bell and Pavitt 1993; Fagerberg and Godinho 2005). Whatever the reasons for this highly uneven innovation system building across developing countries, it is clear that no one set of technology-specific policies is likely to work in every context (Chaminade *et al.* 2009). Not only do developing countries have different histories, legacies, endowments, environments, cultures and politics, but they also face a wide variety of context-specific development challenges and priorities. For example, the many sparsely populated countries of Sub-Saharan Africa face huge challenges in reaching rural communities with roads and electricity grid extensions, compared with more densely populated regions in the world. The livelihoods available to those living in mountainous areas are different to those on the dry savannah, with consequent differences in opportunities to innovate. Countries with particularly high populations of poor people do not provide attractive investment opportunities for business in consumer goods and so it is difficult to create spaces in which to enhance related innovation capabilities.

The extent to which different types of knowledge and technology are likely to be appropriate also depends on a range of context-specificities, such as their applicability in different physical, cultural, political and economic contexts. For example, the technological needs of communities with different wealth levels should be understood – poorer communities, for example, might have a greater need for technologies related to subsistence, while wealthier communities might have priorities around transport or processing goods to add value. This prompts consideration of the extent to which technology transfers facilitated under existing international policy mechanisms are pro-poor (for a useful point of departure, see Urban and Sumner 2009; and Byrne *et al.* 2014b). In poor rural areas, for example, it might be more viable to explore adaptive innovation for low maintenance configurations of solar PV and LED technologies, as opposed to clean options for centralised energy generation that might better suit urban industrial interests. And in adapting to climate change, technologies such as drought-resistant strains of crops, or knowledge regarding new farming methods in increasingly flood-prone areas, might be of more relevance to poor people than advanced engineering solutions to strengthen coastal flood defences.

This concern with the extent to which the needs of poor people are being met through policy interventions around sustainability-related technologies speaks to an emerging interest among development practitioners and researchers the idea of 'inclusive innovation' (see, e.g. IDRC 2011), which ostensibly deals with the extent to which technology innovation and diffusion serve the needs of poor and marginalised people. As Foster and Heeks (2013) demonstrate, while an Innovation Systems perspective is well suited to better understanding innovation in a pro-poor context, this emphasis on inclusivity and pro-poor innovation and diffusion of technologies requires attention to a number of specific considerations that are underplayed in traditional Innovation Systems-based approaches to analysing policy and practice. These include a need to attend to processes of technology diffusion, informal demand-side actors and intermediaries, and the role of local and informal institutions.

A range of different physical and environmental factors are also likely to come into play in determining the context-specific considerations to which policy and practice must attend if useful knowledge flows and capacity building are to be brokered. For example, different wind technology solutions are viable under different ambient conditions, and in other conditions are not viable at all. Carbon capture and storage (CCS) technologies would need to be adapted to suit both local fuel sources and geological storage options (Tomlinson *et al.* 2008). And these physical spatial variations are also likely to play out simultaneously in the form of socio-cultural considerations, e.g. energy efficiency or clean, decentralised energy options need to work within the context of existing cultural (behavioural) practices and existing infrastructure, and so on. So a range of different spatial and socio-cultural factors comes into play when considering what types of knowledge flows and technologies are likely to work or be appropriate in different developing country contexts.

There are also critical context-specific considerations regarding the ways in which knowledge flows are likely to be most effectively brokered in order to build technological capabilities in different developing country contexts. The needs of rapidly emerging economies, for example, are likely to differ significantly from the needs of other developing countries – particularly least-developed countries – in this respect. However, it is important to note that even across such contexts, appropriate levels of knowledge flows are likely to vary according to the specific technology in question and the availability of existing (or related) technological capabilities in different country contexts. There is a need to understand and chart the existing distribution, nature and level of technological capabilities for working with the variety of technologies across and within different contexts. For example, to what extent do different developing countries, or the regions, firms, or communities therein, have the capabilities to work with technologies at different stages of commercial development (e.g. dealing with higher investor risk at earlier stages of technology development), or to work with the hardware and software components involved? One example of this would be a technology like CCS, which involves more complex systems management capabilities than small-scale solar PV (Ockwell *et al.* 2010b).

Consideration of the existing levels of relevant technological capabilities has material implications for which part of the innovation chain would benefit from targeted interventions. So, for example, in Kenya, where solar PV assembly has only recently begun, interventions focussed on the demonstration of process manufacturing techniques might be most appropriate. In the wind industry in China, on the other hand, were it not already considered sufficiently advanced, knowledge flows might be more effectively targeted at international collaborative efforts in R&D (see Ockwell *et al.* 2015 for a discussion of collaborative R&D and climate technology transfer). Such nuanced understandings of relative technological capabilities inter- and intra-nationally have a key contribution to make in better orienting international policy efforts in ways that can be effectively targeted towards nurturing innovation systems and developing technological capabilities.

In their discussion of collaborative R&D and climate technology transfer, Ockwell *et al.* (2015) also draw on Sagar's (2009) typology of the different climate technology needs against which collaborative R&D efforts might be targeted via national or multilateral actions under the UNFCCC. This typology broadly classifies climate technologies into three categories: (1) those that already exist and might meet developing country needs; (2) those that do not yet exist, but which might be developed through targeted policy incentives to meet nascent needs; and (3) technologies that might be required to meet future needs. The discussion above highlights the importance of extending Ockwell *et al.*'s (2015) analysis in order to further consider the appropriateness of such a framework at different stages along the innovation chain, moving away from a fixation on R&D and recognising the potential value (depending on context) of interventions at other stages. As the discussion above has emphasised, authors like Bell (e.g. 2009) have argued that technological capabilities are often developed incrementally in developing country firms, using international knowledge flows that facilitate gradual increases in levels of sophistication. For many countries, and especially the least developed, this implies that international knowledge flows might be much better targeted at climate technologies which are already commercially available, and that technological capabilities be built 'upwards' from there (as in the example of Lighting Africa in Chapter 5). The impact in terms of building sustainable innovation systems in the long term is likely to be no less profound. Note that this also speaks to the importance of policy interventions that attend to the existing levels of capabilities within specific country and technology contexts.

Conclusion: Innovation Systems insights and gaps

In this chapter, we have covered several key insights from the Innovation Studies literature. Taking stock of the discussion above, we now have a more holistic understanding of what constitutes innovation: it refers more broadly to processes and outcomes of change, and to change that can be as much incremental and adaptive as it ever is radical. This gives us important indications of the ways in which innovation might be defined and measured. The continuing emphasis on metrics such as R&D and patenting are demonstrably flawed in their ability to act as meaningful proxies for innovation. The knowledge-based nature of technology and the mostly tacit nature of this knowledge are also revealing, especially when looking to understand how developing country firms have, in the past, been able to accrue technological capabilities, underpinning economic development trajectories that have led to the process of 'catching up' in countries such as the Asian Tigers. When thinking about broad ideas like 'low carbon development', or more specifically about the issue of sustainable energy access, such insights give us essential indications of the likely nature of technological capability development and the underpinning knowledge transfer we might expect to drive the processes of innovation and technological change that such ideas imply for developing countries. And the positioning of these processes within the broader contexts of countries' innovation systems provides us with important purchase in terms of focussing international and national policy efforts. That is, these efforts must shift away from 'Hardware Financing' and 'Private Sector Entrepreneurship' framing, and towards creating the systemic contexts (fertile gardens) within which technological change can be nurtured and sustained. Finally, the discussion of context-specificities emphasises the differences in how and where international policy and practice efforts might be focussed when driving transformations in access to sustainable energy technologies in developing countries.

To some extent, this focus on context-specificities might be interpreted as con-founding policy efforts, especially those at the national and multilateral levels such as SE4All and the UNFCCC. Some observers might push instead for the identification of non-context-specific issues so that more generic policy approaches can be developed and applied. Indeed, there has been, and still is, a tendency for international climate policy to focus at this level. For example, the generic failure of markets to capture the positive externalities of lower carbon technologies was an important rationale behind the kind of 'Hardware Financing' approach that has characterised policy in this field to date (e.g. the CDM, as discussed in Chapter 1). Other generic issues might include lock-in (Unruh 2000) to high carbon technologies, or to agricultural technologies that are over-reliant on high levels of water or fertiliser inputs. However, notwithstanding country-driven activities through institutions such as the Global Environment Facility (GEF), the tendency to focus on policy options that are not sensitive to context-specificities is, as argued above and later in this book, a major reason why many sustainability-related technology efforts under international mechanisms such as the UNFCCC have met with limited success to date. As we expand upon in detail in this book, a policy focus on nurturing innovation systems is the way in which these past policy shortcomings can be overcome. Interventions that aim to play an 'innovation system builder' role in developing countries provide the basis for designing generic policy approaches that simultaneously respond to context-specificities.

However, as much as the above insights from Innovation Studies are important and revealing, there remains much that needs to be questioned and elaborated within the context of an interest in sustainable energy access in (particularly low-income) developing country contexts. The first shortcoming of the Innovation Studies literature is that it has mostly been concerned with either OECD countries or the more recent rapidly developing economies. Relatively little of the theory has been developed on the basis of the empirical context of low-income countries which, as noted in Chapter 1, differ in important ways from any semblance of the 'modern infrastructure ideal' (Furlong 2014). And while a nascent literature exists to apply Innovation Studies approaches to deal specifically with low carbon technologies, there is still a need for further empirical work to develop these perspectives in relation to the specific challenges that low carbon technologies raise. These include dealing with conditions of urgency in relation to climate action, dealing with technologies that in some cases are commercially immature, and dealing with the failure of the market to capture the positive externalities of carbon mitigation (Ockwell and Mallett 2012) as well as, potentially, poverty reduction.

But, as well as this need for more testing and development of Innovation Studies approaches within specific empirical contexts and in relation to particular issues, we would argue that there are also theoretical weaknesses in regard to the problem of sustainable energy access. To some extent, as the discussion of innovation processes above demonstrates, the literature has considered technology users only in terms of the knowledge feedbacks firms receive from marketing and sales activities. This is arguably a rather limited perspective on the role of technology users and society more broadly. The Innovation Studies literature itself recognises these contextual factors as important for motivating the innovation activities of firms. But, it has not gone far in seeking to theorise the social contexts within which new technologies might be used, nor in seeking to theorise how sustainable energy technologies might have to compete with the incumbents that underpin the social practices that energy facilitates. And yet any transformation in access to sustainable energy technologies implies that such social contexts could be both transformative for, and transformed by, these technologies. As emphasised early on in this chapter, innovation is not just about technological invention; 'it involves change of many kinds: cultural, organisational and behavioural as well as technological' (Stirling 2015b, p. 1). In this sense, the now well-established literature on Socio-Technical Transitions is far better positioned to contribute. In Chapter 3, we therefore discuss the insights this literature has to offer in relation to the problem of sustainable energy access and how a theoretical framework based on an innovation systems perspective might be extended accordingly. This takes us beyond the firm- and industry-centric focus of the Innovation Studies literature to incorporate the context of poor people and the social practices with which sustainable energy access intersects. In this way we are also able to return to Stirling's (2015b) definition of innovation as one that goes beyond the technology-centric way in which the term has been used in this chapter.

Note

1 Cohen and Levinthal (1990, p. 128) specify absorptive capacity as the ability of a firm to 'recognize the value of new information, assimilate it, and apply it to commercial ends'.

3

INNOVATION IN THE CONTEXT OF SOCIAL PRACTICES AND SOCIO-TECHNICAL REGIMES

Introduction

As we have argued in Chapter 2, a systemic perspective on the context within which innovation and technological change occur provides us with significant analytic purchase, particularly in terms of where interventions in policy and practice might usefully focus in seeking to transform specific technological contexts, such as the use of sustainable energy technologies in developing countries. In this chapter, we move beyond the firm-centric focus of the Innovation Studies literature to consider the social contexts within which it is assumed these sustainable energy technologies will be used, namely, the everyday lived experiences of energy consumption by poor people in developing countries. As argued at the end of Chapter 2, the simple treatment in the Innovation Studies literature of technology users as providing knowledge feedback mechanisms into firms' innovation processes tells us little of the social contexts within which sustainable energy technologies might be used, or of the incumbent energy technologies and associated social practices with which sustainable energy technologies must compete. In this chapter, we therefore extend our account of innovation and technological change to engage with insights from the literature on Socio-Technical Transitions and, more specifically, the stream of literature on Strategic Niche Management. This offers us a more holistic perspective on processes of (socio-)technological change and innovation by paying analytical attention to non-sustainable incumbent technologies, technology users and the social practices concerning energy with which sustainable technologies must necessarily intersect. We conclude this chapter by articulating a broader theoretical perspective based on building socio-technical innovation systems as the means through which transformations in sustainable energy access might be achieved.

The socio-technical nature of innovation and technological change

With its intellectual roots in evolutionary economics, and using 'insights from historical, sociological and actor-network studies' (Raven 2005, p. 25), as well as contributions from anthropology and, more recently, political science (e.g. Kern 2011; Meadowcroft 2011) and geography (e.g. Lawhon and Murphy 2012), the rapidly growing body of work on socio-technical transitions (e.g. see Rip and Kemp 1998; Geels 2002; Berkhout et al. 2004; Raven 2005; Geels and Schot 2007; Smith and Stirling 2007; Smith et al. 2010; Byrne 2011) understands a technology to be more than a discrete artefact. The artefact (piece of technology hardware) is perhaps the most visible part of a 'technology' but it is embedded within a complex social, economic, technical and political system (although the political dimension of the system is arguably the least well-developed part of Transitions theory to date, e.g. see Smith and Stirling 2010; Meadowcroft 2011). This perspective facilitates an explicit consideration of the technology user and the social practices with which technologies intersect. It facilitates attention to several key issues of relevance to the adoption of sustainable energy technologies by poor people in developing countries – considerations that are not addressed by mainstream policy thinking at present.

Definitions of socio-technical systems vary to some extent, but this system is conceptualised as a technological regime, described by Hoogma et al. (2002, p. 19) as 'the whole complex of scientific knowledge, engineering practices, production process technologies, product characteristics, skills and procedures, established user needs, regulatory requirements, institutions and infrastructures'. This notion implies path dependency, in that current knowledge, practice and so on depend on what has gone before; and a technological trajectory, in that current knowledge, practice and so on guide, but do not determine, what will happen in the future (Nelson and Winter 1982, pp. 262–263; Dosi 1988, p. 225; Dosi and Nelson 1993, p. 30).

The regime is able to exist because it serves some societal need,[1] such as electricity supply,[2] in a way that is acceptable or convenient to that society. In other words, there is an important interdependent relationship between a 'technology' and the wider social context such that each can effect change in the other. The Transitions literature provides us with windows into various ways in which marginal, experimental or sometimes radical socio-technical practices can come to influence mainstream practices and even thoroughly transform them over time (Geels and Schot 2007). Some regimes can change completely: they are organised around radically different technological artefacts with different bases of scientific knowledge, institutions and so on, but servicing familiar needs. Some are completely new regimes: previously unknown possibilities are realised, such as the development of passenger air travel. It is these socio-technical transformations that are the focus of the Transitions literature – analytically, how and why they happen; and, normatively, how they can be realised.

Developments in the literature in the last decade or so have resulted in a now widely used framework for organising the complex relationships in socio-technical systems (see Figure 3.1). Socio-technical regimes are conceptualised at a meso-level,

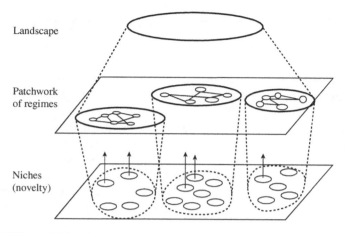

FIGURE 3.1 The multi-level perspective
Source: Geels (2002, p. 1261).

with a landscape at the macro-level and niches at the micro-level, to form a multi-level perspective (Geels 2002).

The landscape refers to a heterogeneous set of factors such as economic growth, war, cultural norms, etc., where the defining feature of these factors is perhaps best understood in terms of their being outside the direct influence of human actors (Romijn *et al.* 2010), and socio-technical niches refer to 'proto-regimes' where novel technologies are the focus of experimentation and learning (Geels 2002, pp. 1259–1261). Transitions are then defined as 'changes from one sociotechnical regime to another' (Geels and Schot 2007, p. 399). A key feature of a socio-technical regime is that trajectories from different domains – policy, science, technology, culture, users and markets – are aligned; they are mutually reinforcing (see Figure 3.2), so that 'to understand dynamics in [socio-technical] systems we should look at the co-evolution of multiple trajectories' (Geels 2004, pp. 911–912).

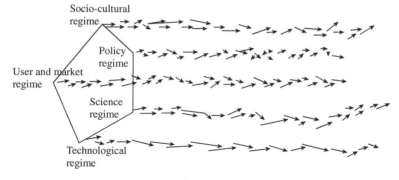

FIGURE 3.2 Alignment of regime trajectories
Source: Geels (2004, p. 912).

Socio-technical transitions and sustainable energy access

Of specific interest to our focus on sustainable energy access, but also of relevance to understanding low carbon development more broadly, a socio-technical perspective allows us to understand technologies as co-evolving with the social contexts within which they are used. For instance, new technologies will be widely adopted not simply because they successfully harness technical principles but also if their form and function are aligned or 'fit' with dominant social practices, or offer opportunities to realise new practices – to 'stretch' existing socio-technical practices – that are attractive in particular social and geographical settings (Hoogma 2000; Raven 2007). Innovation and technological change are thus understood as processes that occur within the context of social practices, shaping – and shaped by – such practices and evolving local knowledge. Examples of local knowledge include the cultural practices in cooking or expectations about personal mobility, each providing specific conditions within which low carbon energy alternatives might 'fit' with, or from which they might 'stretch' beyond, current practice. This allows analysis to focus on the services for which electricity is desired or used (e.g. lighting or mobile phone charging), and so better understand how and why people might use any specific technology that facilitates such services (e.g. a solar home system or a solar lantern). Our attention is, therefore, drawn to analysing the multiple context-specific dynamics within which it is assumed sustainable energy technologies will be used. Furthermore, as well as attending to existing institutions and market structures (e.g. supply chains for kerosene and subsidies intended to help the poor), a socio-technical perspective directs us to analyse how the dominant nature of such institutions creates a powerful inertia against – in our case – low carbon energy alternatives.

All these ideas are potentially useful for enhancing our understanding of large-scale or long-run socio-technical change but they have been developed mainly in the context of highly industrialised economies. As a result, there are challenges in operationalising a Socio-Technical Transitions perspective in relation to sustainable energy access among poor people in developing countries, and especially in the many countries categorised as low income. The landscape and niche constructs are relatively straightforward to apply – at least superficially – but the regime concept is much more problematic, especially in the case of rural electrification using sustainable energy technologies. This is not to say that regimes are unidentifiable in these contexts; rather, it is to say that such identification remains an empirical challenge requiring careful analytic attention.

The core focus of the Strategic Niche Management (SNM or niche theory) stream of Transitions literature, for example, is on understanding sustainable energy technologies as part of a sustainable 'niche', or protected space, in which the normal selection pressures that help the dominant fossil fuel-based energy 'regime' to reproduce itself are weakened or absent (Smith 2007; Smith and Stirling 2007). As outlined above, a socio-technical regime includes incumbent technologies as well as established values and practices that are socially embedded and that follow an established pathway that reinforces the current stable technological system:

'incumbent systems, such as large-scale, centralised, fossil-fuel electricity generation, constitute more structured and structuring "socio-technical regimes"' (Smith *et al.* 2014, p. 117). In the context of many developing countries and many sustainable energy technologies, any large-scale stable system is difficult to identify. For example, off-grid solar PV-generated electricity, as a provider of lighting and phone charging, could replace kerosene and batteries or other fossil fuel-powered charging devices. The currently common practices of lighting and phone charging can be conceptualised as the regime, and the off-grid solar PV niche either competes with it or seeks to significantly reduce its dominance. It is not the same kind of stable large-scale regime that Socio-Technical Transitions studies traditionally incorporate, but it seems to be a suitable way to frame the regime in the case of the off-grid solar PV socio-technical niche, as any large-scale incumbent technology is absent. On the face of it, a regime characterised by kerosene and batteries might not seem as pervasive as, say, coal-fired and grid-based electricity provision in an urban industrialised country context. Nevertheless, the chain of power dynamics and vested interests pertaining to the supply of kerosene or batteries to rural communities in developing countries should not be underestimated. Nor is the adoption of low carbon technologies for providing the same energy services any less exposed to the kinds of challenges that niche technologies face in many contexts when seeking to influence or compete with existing regimes. As outlined above, Geels' (2002) multi-level perspective (MLP) also posits the existence of a 'socio-technical landscape' over and above the regime, constituted by exogenous factors that can put pressure on the regime and open up windows of opportunity for niche configurations to break through (Markard *et al.* 2012, p. 957). In the case of solar lighting and phone charging, external influences might include climate change and health concerns regarding kerosene and fossil fuel-based electricity, as well as the need for poverty alleviation, especially energy poverty alleviation. All these aspects are motives for clean and affordable electrification solutions.

Notwithstanding the challenges above, a socio-technical perspective therefore provides us with the tools to understand how processes of sustainable energy technology adoption among firms and users might happen, how such actors participate in technology development and innovation, and how policy might pursue systemic interventions to nurture such change. This includes, for instance, understanding examples of sustainable energy technology projects in developing countries as 'experiments' conducted in a 'protective space' (niche) that enjoys freedom from competition with the dominant technologies in the regime, e.g. unfamiliar low carbon energy options and practices competing with familiar fossil fuel-based energy systems, with their attendant social norms and powerful vested interests, as well as existing infrastructure and established technologies. This casts light on the way a more systemic policy approach – focussed on nurturing socio-technical innovation systems – can serve to connect and learn across sustainable energy technology experiments and niches through strategic interventions, whether at firm, industry or user levels. As noted above, the SNM literature attends to these niche-regime dynamics within the context of transitions towards sustainable socio-technical regimes, and so we take some time now to elaborate its main concepts. This provides the basis upon which we are able to operationalise Socio-Technical Transitions theory as we use it for the historical analysis of the off-grid solar PV market in Kenya in subsequent chapters of this book.

Operationalising a socio-technical transitions perspective

Strategic Niche Management is a framework that can be applied either analytically or normatively (Raven 2005, p. 37). It has been used both as a management tool to design policies and experiments for socio-technical niches (see Caniëls and Romijn 2008; Schot and Geels 2008) and, as we use it in the empirical analysis in this book, a theory to analyse niche processes *ex post* (although note that our intention is for the outcome of this *ex-post* analysis to inform future policy design – in this sense we are therefore using SNM both analytically and normatively). Through such *ex-post* analysis, sustainable energy technologies and the energy services they provide, along with associated actor networks and relevant institutions, can be conceptualised as a *socio-technical niche*.

In its normative mode, SNM theory is intended to be used to find and develop sustainable solutions to societal needs. Central to the approach is the creation of experiments with promising technologies in social contexts in order to generate opportunities for co-evolutionary learning. As promising technologies are often unready to face real-world selection pressures, however, their experimental application in social contexts requires some form of protection (Schot et al. 1994; Kemp et al. 1998; Raven 2005; Schot and Geels 2007; Smith and Raven 2010). This 'protection' is the essence of the niche concept: the niche is a real-world space in which normal selection pressures – such as competitive price and performance characteristics – are suspended, weakened or changed in order to enable the survival of the promising technology so that actors can learn about its desirability and develop it further (Kemp et al. 1998). Selection pressures can be changed in a number of ways, for example, through the use of public subsidies to bring costs in line with those of incumbent technologies; or R&D budgets can be used to support real-world trials or demonstration systems, where there is no requirement for any direct commercial return; or new regulations can be introduced to favour the promising technology over the incumbent (Schot et al. 1994; Schot and Geels 2007). Implicit in the notion of 'promising technology' is an expectation or belief by actors that the technology will eventually be socially and commercially viable, and so they are willing to protect it and invest effort to develop it (van Lente 1993; Raven 2005; Schot and Geels 2007). The hope, or objective, is that the attempt to develop the technology (and adjust the system in which it is embedded) will indeed lead to a socially and commercially viable innovation so that the protection can be removed and a new socio-technical regime will emerge based on the innovation (Schot et al. 1994; Kemp et al. 1998; Raven 2005).

Proponents of SNM suggest that a niche innovation is more likely to become an element of a new socio-technical regime when niche experiments are rich in 'second-order' learning and carried out by a broad network of actors, including users (Hoogma *et al.* 2002, p. 194). In contrast to 'first-order' learning, which is concerned with the functioning of a technology (important though that is), second-order learning arises 'when conceptions about technology, user demands, and regulations are … questioned and explored' (Hoogma *et al.* 2002, p. 29). In other words, second-order learning occurs when assumptions and behaviour patterns are examined in conjunction with the use of particular technologies. A broad network of actors is important to generate lessons about product integration, admissibility and acceptance, which are processes that Deuten *et al.* (1997, p. 132) describe as 'societal embedding'.

But this discussion is concerned with SNM in its normative mode: that is, where SNM is used explicitly as a method for finding and developing sustainable solutions to societal needs. In this mode, a niche – a protective space – would be *created* by implementing various measures to change selection pressures, and experiments in social contexts would be designed to stimulate first- and second-order learning. After some time, the protection would be removed and there would be a hope of widespread adoption of the niche innovation. This clearly raises considerable promise in relation to interventions through policy and practice to transform sustainable energy access. As emphasised in Chapter 1, however, understanding the extent to which such transformations are possible, and the nature and characteristics of the kinds of actions that might drive them, demands careful analytic attention to existing examples where such transformations may have been effected. In these cases, if we wish to learn something about the introduction and development of new (sustainable energy) technologies in practice, we would need to use SNM analytically. We then need to empirically identify the various categories and concepts of SNM in operation. These categories (discussed in more detail below) consist of: a protective space or niche; experiments; relevant actor networks; expectations and beliefs about promising technologies; first- and second-order learning; the enabling and constraining institutional environment, including social norms, professional practices and formal institutions; and the context in which the innovation is situated. All these interact dynamically over time and so we need to trace them and their interactions historically. SNM, therefore, focusses our attention on the niche of the multi-level perspective (MLP) and a set of dimensions that links with Geels' (2004, p. 912) multiple trajectories. So, in its analytical mode, SNM directs us to examine: (1) the processes and quality of learning; and (2) the composition and quality of social networks as they relate to technological experiments in a social context.

Having emerged from research in industrialised-country contexts, the SNM approach has, in recent years, begun to be applied in the context of developing countries. For example, see the special edition of *Environmental Science and Policy* introduced by Berkhout *et al.* (2010) for the application of these ideas to developing Asia, and see Byrne (2011), Byrne *et al.* (2014b) and Rolffs *et al.* (2015) for their application in Kenya and Tanzania. These papers focus on the use of niche theory to understand the dynamics of how novel technologies were tested in real-world settings, and to identify the ways in which they resulted in wider use and further development. A better understanding of these dynamics and processes raises the potential to better design how policy could deliberately intervene to nurture sustainable niches. A policy might aim, for example, to widen and deepen access to low carbon energy technologies to benefit poor and marginalised groups, and do this by creating new – or nurturing existing – niches of low carbon energy technology applications among poor communities and households. Important in this regard, is that some niche analyses reveal the role that key actors, such as 'cosmopolitan actors' (Deuten 2003), or what we refer to in this book as *socio-technical innovation system builders*, can play in developing a niche. It is in understanding the role of such actors, most specifically, that we can find inspiration for policy interventions. That is, policy-makers and other actors (e.g. NGOs or private companies) could learn from the actions of past successful innovation system builders, and apply this learning through interventions that seek the increasing adoption of sustainable energy technologies and the achievement of wider sustainable development impacts.

As well as articulating a conceptual framework capable of matching the ambitions of international policy, the key aim of this book is to provide a rich historical account of one example of transformative change in relation to sustainable energy access. Within the broader context of a socio-technical innovation systems perspective – which we articulate in more detail below – we therefore use several concepts from SNM to assist us in our historical analysis. We have already mentioned these concepts in the preceding discussion but we now elaborate them in some detail.

Protective space

SNM argues that sustainable innovations need 'protective spaces' where experimentation and the development of new technologies can take place within a supportive environment (Smith *et al.* 2014). Jacobsson *et al.* (2004, p. 24) emphasise that spaces offer opportunities for learning and that their protection goes beyond technology policy instruments. Protection is essential at the initial stage of innovations as it 'shields' them from mainstream selection pressures. Shielding can be passive through the use of pre-existing configurations, or active through strategic intervention from actors. The process of 'nurturing' enables socio-technical niches to grow and become able to influence or enter the regime. Nurturing is defined as support processes that enable the development of innovations. Dedicated intermediary work is needed for interactive learning to take place, expectations to develop, and supportive networks to build (Smith 2007). After that stage, the protection of the niche shifts to 'empowerment'. Empowerment is mainly achieved by advocates and networks around the socio-technical niche. Actors within the socio-technical niche become outward-oriented, meaning they are active within the regime and interact with others (Smith and Raven 2012; Smith *et al.* 2014, p. 117). The subsequent analytical categories within SNM are embedded within the protective space, but constitute important additional foci because they force our attention towards the dynamics of interacting change processes.

Experiments and learning

Experiments can be perceived as being part of the process of nurturing (Smith *et al.* 2014, p. 118). They are defined as 'initiatives that embody a highly novel socio-technical configuration likely to lead to substantial sustainability gains' (Berkhout *et al.* 2010, p. 262). Experiments can be 'local', which means that they take place in specific places, supported through local networks. They generate lessons that lead to learning (Smith and Raven 2012). For example, in relation to off-grid solar PV in Kenya, finance projects, programmes and the business models of socially oriented enterprises have been experimenting with the provision of end use-level finance. These experiments might therefore be interpreted as potentially generating learning that could strengthen the off-grid solar PV niche. As mentioned above, SNM conceptualises two types of learning that are important for us to analyse when investigating niche processes: first- and second-order learning. Below, we provide an elaboration of these concepts, as SNM posits them.

First-order learning

Hoogma *et al.* (2002) define first-order learning as the testing of a technological artefact, or the technical configuration of artefacts. The motivation here is to understand how to make that *particular* artefact or configuration work, rather than to explore alternatives. In other words, an experiment begins with a prospective technological solution rather than the intention to investigate a problem. Consequently, users are not challenged to question or explore their needs; there are few opportunities to open up or discover novel approaches, and so there is little opportunity for co-evolutionary learning to occur. Many valuable lessons can be generated by first-order learning but they will be concerned with 'how to improve the design, which features of the design are acceptable for users, and about ways of creating a set of policy incentives which accommodate adoption' (Hoogma *et al.* 2002, p. 28). First-order learning is therefore instrumental, focussed on trying to make a particular socio-technical configuration work. It is concerned with refinements to the particular socio-technical configuration and tends to result in the accumulation of facts and data. For example, these could be about the technical performance or characteristics of a specific technology or finance model and ways in which these characteristics could be improved.

Second-order learning

Second-order learning is more fundamental. It is understood to contrast with first-order learning because it is about investigating the assumptions concerning a particular societal function: that is, for example, the assumptions around mobility or communication in a particular society. By doing so, SNM posits that 'co-evolutionary dynamics' will arise, whereby there will be 'mutual articulation and interaction of technological choices, demand and possible regulatory options' (Hoogma *et al.*

2002, p. 29). In other words, second-order learning is thought to be fundamental learning about the function itself within its social context, not simply a kind of instrumental learning about a particular technological solution. It can occur when the framing assumptions of a particular socio-technical configuration are challenged and can therefore result in a new set of framing assumptions and a new config-uration. This new configuration will then require further first-order learning to accumulate relevant facts and data, and to establish working refinements. To the extent that niches are protective spaces in which to experiment, learning can generate a range of socio-technical configurations and each can develop in parallel with the others. But, given the experimental nature of niches and the uncertainties associated with any particular socio-technical configuration, it is likely that second-order learning is especially critical to developing configurations that work and that can be successfully – and widely – deployed (Schot and Geels 2008; Byrne 2011).

Articulating learning

Learning in itself, whether first- or second-order, is not enough to have an impact on behaviour and change. It must be articulated and disseminated if the lessons are to provide useful information with which to make adjustments to behaviour or interventions (Raven 2005, p. 42, following Ayas 1996, p. 39). Raven elaborates this further, explaining five 'methods' of learning that may be ineffective, as Ayas identifies them. These are: (1) role-constrained (an actor's role prevents further action); (2) superstitious (inadequate reasoning or basis for changed behaviour); (3) situational (uncodified learning); (4) fragmented (learning is not disseminated); and (5) opportunistic (actors learn something useful but lose influence). So, if it is to be useful, the implications are that learning should be interpreted with an understanding of who is generating the lessons, the evidential basis for the lessons, and that it needs to be codified and disseminated.

Actor networks

Niche experiments involve actors who interact through networks. Networks of actors are important to build robust support for socio-technical practices, to facilitate knowledge exchange, to enable interactions between stakeholders, and to provide access to resources. SNM posits that the composition of these networks is important for access to resources and for the opportunities created – or constraints imposed – by the diversity or otherwise of network participants (Raven 2005, pp. 39–41). Networks might be more effective if they are broad, which means the involvement of a large variety of stakeholders, and if they are deep, meaning there exists strong commitment among all actors and organisations (Schot and Geels 2008, pp. 540–541). A narrow network is one that is dominated by regime 'insiders' who are, it is assumed, more likely to maintain the status quo or favour only incremental innovations, than to support radical solutions to a function's needs. A broad network, by contrast, would include regime 'outsiders', users and non-users[3] (Hoogma *et al.* 2002, pp. 192, 194). However, it is not enough for a broad range of actors to be connected; the quality of their interactions is important. That is, the network members should be *active* participants in the innovation process (Raven 2005, p. 40).

Actors may be attracted to a network through the deployment of expectations. However, this does not mean that their expectations are necessarily well aligned with others in the network; there may still be work to do in aligning expectations (Raven 2005, pp. 40–41). The point of this is to establish the shared rules and so on, that Geels and Raven (2006) identify as important for setting and consolidating socio-technical trajectories. Moreover, the alignment of actors in this way helps to build a constituency of support around a particular solution (Kemp *et al.* 1998, p. 186), and stabilise actor relations (Hoogma *et al.* 2002, p. 29). Finally, apart from this network alignment, the work of actors needs to be directed at network integration – that is, the forming of ties to networks outside the niche, such as those of suppliers of complementary technologies and strategic elements of the regime (Deuten *et al.* 1997, p. 132; Weber *et al.* 1999, p. 18).

Expectations and visions

In SNM, expectations and visions are general articulations of the future in which particular socio-technical configurations are usually central (Byrne 2011). For example, rural electrification based on SHSs can be considered an expectation. A vision would include more than this general articulation by including the means by which the expectation can be realised. Such means might include business models, supportive policies and technical specifications for the SHSs themselves. Both expectations and visions can be linked directly with first- and second-order learning. That is, an expectation can act as a goal, arising from a set of framing assumptions, towards which actors engage in first-order learning as they try to realise the expectation. In doing so, they begin to detail the means by which that expectation can be realised and therefore begin to detail a particular vision. When framing assumptions change through second-order learning, a new expectation is generated and further first-order learning in this new direction will begin to detail a new vision. However, both expectations and visions need to be sufficiently robust, specific and stringent – and be 'shared' collectively – to have long-term effects on the evolution of a niche (Raven 2005).

Recent work in the literature has attempted to develop these ideas, including an attempt to bring precision to their definitions. As a result, it is apparent that there is a range of understandings of the meaning of expectations and visions. Perhaps two of the clearest definitions attempted are those by Berkhout (2006) and Eames *et al.* (2006). The first of these, given by Berkhout (2006, p. 302), proposes a definition of visions as 'collectively held and communicable schemata that represent future objectives and express the means by which these objectives will be realised', while Eames *et al.* (2006, pp. 361–362) suggest 'visions … refer to internally coherent pictures of alternative future worlds. Normative in character, visions are explicitly intended to guide long-term action. … expectations … refer to less formalised, often fragmented and partial, beliefs about the future.'

Thus, expectations and visions are conceived to be cognitive representations of a technological future, which either guide action or articulate what action is required in the realisation of that future. Furthermore, as Berkhout implies, we can distinguish between *individual* and *collective* expectations and visions. In general, collective expectations are those that are of interest to the analyst because individual expectations are 'not likely to be socially significant, even [when] held by a powerful social actor' (Berkhout 2006, p. 301). Geels and Raven (2006, p. 375) are less explicit in this sense but they imply a similar interest when they talk of *shared* routines and their effect on local actors' practices, the significance being that when 'technical search activities in different locations are focused in a similar direction, they add up to a technical trajectory'. The implication of this is clear when considering the role of expectations in socio-technical niche development processes: a socio-technical trajectory may lead to the emergence of a new socio-technical regime.

Perhaps the most fundamental function of expectations is already expressed in their definitions. As van Lente (2000, p. 43) says: 'One of the striking things about technological futures is that they often appear in the imperative mode. That is, once defined as promise, action is required.' This is supported by Michael (2000) when he suggests that the way in which expectations and visions are expressed does rhetorical work; the articulation of a particular technological future is not neutral. In Berkhout's (2006, p. 300) terms, visions are 'encoded or decoded as either utopias or dystopias'; they are 'moralised' in order to enrol actors because the benefits and dis-benefits of particular futures will be unevenly distributed. Of course, to identify a strategy is not to explain the enrolment of others. As Konrad (2006, p. 432) notes: 'The analytical focus on the coupling of interests and expectations does not explain why others should accept these promises, especially if they have no strategic interest in the technology in question.' Konrad's point is to caution against a unidirectional understanding of expectation dynamics. Certainly, the rhetorical force of visions of the future is important but these visions emerge from *interactive* processes – through public and specialist discourses and innovation activities. As such, Berkhout (2006, pp. 301–302) argues that:

> It may be more productive to see expectations as 'bids' about what the future might be like, that are offered by agents in the context of other expectation bids. Expectations offer a potentiality that in most circumstances requires the endorsement and affiliation of other actors before it can be actualised.

Drawing on this discussion, we can say that expectations are, in essence, motivators of action in a particular socio-technical direction. That is, they operate as targets towards which actors can align themselves and their activities. Visions are more detailed; they specify the means to achieve the target. But different actors create different targets and so there is always a process of negotiating the particular contents of expectations when, as is necessary, the attempt is made to recruit resources. Always at stake, of course, is the uneven distribution of benefits and dis-benefits that any realised expectation or vision would bring, and whether this would mean a change from the present (uneven) distribution. In other words, the process of creating collective expectations and visions is inherently political: it involves strategic interest, power and authority; persuasion, negotiation and conflict.

Institutions (and institutionalisation)

SNM sees institutions as playing an important role in the societal embedding of niche technologies. Institutions here are understood in their broadest sense to include laws, regulations and policies as well as practices, norms and conventions regarding a particular socio-technical configuration (Byrne 2011, p. 19). A critical process in developing from a niche to a regime is the structuring of practices that can be widely adopted. Institution-building (whether of formal or non-formal institutions) is therefore an important process that co-evolves with those outlined above as a niche develops. From the perspective of off-grid solar PV in Kenya, for example, this would include the co-evolution of governmental regulations – such as import taxes or quality standards – along with solar technologies.

This category also directs analytical focus towards consumer behaviour, socio-cultural practices and the relevant energy services that sustainable energy technologies might facilitate. To some extent, we have already encountered this in the discussion of 'fit' or what Deuten et al. (1997, p. 132) call 'integration' with the 'existing practices and cultural repertoires of users'. But, following Deuten et al., we could add two other aspects to this: admissibility and acceptance. Both these aspects relate to institutions as we have defined them above; 'admissibility' is about compliance with laws and regulations, some of which may be inappropriate or unavailable for a new technology, while 'acceptance' refers to the perceptions of users and non-users – there may be resistance to a new technology.

The process of embedding these institutions is what we might call structuring, or institutionalisation. Clearly, networks are important for this process as they facilitate the diffusion of institutions (such as best practice, heuristics, and so on) that are developed or adapted through the learning generated in experiments. This helps to structure socio-technical niches and can be further enhanced by the actions of what Deuten (2003) calls 'cosmopolitan actors'. These are actors who work at a level 'above' individual experiments – the cosmopolitan level – and help to spread the lessons from particular experiments to other locations and other experiments. According to Deuten, all technological knowledge begins in a local context and so work must be done to spread that knowledge if it is going to contribute to the emergence of a new technological regime. For this to happen, the knowledge must be made 'translocal'; that is, it must be de-contextualised so that it is available for application by others in new contexts. In turn, application in new contexts creates new knowledge that also must be de-contextualised for further 'transfer'. Cosmopolitan actors coordinate these transfers and, in doing so, increasingly structure local practices; that is, they institutionalise practices. This is not to say that local practices first emerge from a vacuum; there are institutions already in place and these influence practice, but these institutions are adjusted, and changed, and new ones are created, in response to practice.

As noted above, the concept of cosmopolitan actors can provide important insights for policy-makers and practitioners who wish to intervene in order to deliberately drive transformations in sustainable energy access. However, the cosmopolitan actor notion as currently defined in the literature is narrowly concerned with the generation and transfer of codifiable knowledge, and how these processes influence *technical* practices. For our broader interest in socio-technical innovation systems as the contexts within which transformations in sustainable energy access might be nurtured, we need to similarly broaden the definition of the cosmopolitan actor role. In order to capture this broader definition, we therefore refer to such actors as *socio-technical innovation system builders*. As will become clear through the empirical analysis in subsequent chapters, further theoretical articulation of the nature and role of such actors is a key area that this book identifies for future work.

Socio-technical trajectories and systemic interventions

Before we conclude this chapter with an articulation of how the Socio-Technical Transitions perspectives discussed above can be incorporated into this book's broader conceptual framework, we should explain what we mean by a socio-technical trajectory – something we have mentioned in several places along the way. One reason for explaining the concept more fully is that it helps to clarify what we mean by 'systemic intervention', something we have also mentioned several times in our discussions. But we begin the explanation by first considering the narrower idea of a 'technological' trajectory. This notion, unsurprisingly, is closely bound to the concept of technology itself. In Sahal's (1981, p. 22) view, 'technology is as technology does', a functional or, in Sahal's terms, systems view. From this perspective, Sahal suggests that it is possible to measure technical change by observing, for example, changes in power-to-weight ratios, thermal efficiency and so on, and to do so in some objective sense. Dosi (1982) has a similar view but defines two levels: technological paradigm and technological trajectory. Concerning the technological trajectory, Dosi (1982, p. 152) writes: 'We will define a technological trajectory as the pattern of "normal" problem solving activity (i.e. of "progress") on the ground of a technological paradigm.'

And, in regard to the notion of a technological paradigm, Dosi (1982, p. 153) writes:

> The identification of a technological paradigm relates to generic tasks to which it is applied (e.g. amplifying and switching electrical signals), to the material technology it selects (e.g. semiconductors and more specifically silicon), to the physical/chemical properties it exploits (e.g. the 'transistor effect' and 'field effect' of semiconductor materials), to the technological and economic dimensions and trade-offs it focusses upon (e.g. density of the circuits, speed, noise-immunity, dispersion, frequency range, unit costs, etc.). Once given these technological and economic dimensions, it is also possible to obtain, broadly speaking, an idea of 'progress' as the improvement of the trade-offs related to those dimensions.

Using these ideas, we could say that PV systems constitute a technological paradigm. The tasks are to generate and store electricity; the material technologies selected consist of, among others, silicon-based semiconductors, lead acid-based energy storage, electrically-powered lights, and so on; the properties exploited include the photo-effect, the galvanic effect, and so on; and the technological and economic trade-offs include PV cell efficiency, battery charge and discharge characteristics and so on versus production costs. A particular pattern of problem-solving activity within this paradigm would then define a particular technological trajectory. For example, the problems of cell efficiency and production costs have led to the development of amorphous silicon cells; they have much lower cell efficiency than mono or poly-crystalline cells but are much cheaper to manufacture, as they require lower-grade silicon and are more amenable to continuous production techniques (ICCEPT and E4tech 2003). Hence, we could characterise an amorphous-module PV system trajectory alongside others, such as a monocrystalline-module PV system trajectory.

But we are interested in *socio*-technical trajectories. Therefore, we need to expand our definition to include the relevant social dimensions. We are interested in analysing socio-technical *niches* in this book, and we are assuming that a niche trajectory is, in important ways, a nascent regime trajectory. This brings us back to the Hoogma *et al.* (2002, p. 19) definition of a socio-technical regime, repeated here for convenience: 'the whole complex of scientific knowledge, engineering practices, production process technologies, product characteristics, skills and procedures, established user needs, regulatory requirements, institutions and infrastructures'. So, in general, a socio-technical trajectory (whether niche or regime) is a particular pattern of these elements and the focus of problem-solving activities within the pattern. Strictly speaking, changes in any of these elements would constitute a new trajectory. However, there must be room within this view to make a case for whether a change in an element really constitutes a change in trajectory or not.

We can represent this notion of a socio-technical trajectory schematically, in order to give us a clearer view of what is involved (see Figure 3.3). We do not wish to claim that the elaboration given here is a complete description of a socio-technical trajectory. Figure 3.3 may be either too simple or too complicated. The point is to convey the notion of the various 'fronts' on which socio-technical work could be done as a technology and its associated practices evolve. There is also room for further discussion within the idea. For example, 'user needs' are further divided into preferences, practices and requirements. We might reasonably argue that user practices are institutions and so should be classified under that grouping. Indeed, we have operationalised institutions in this way. However, for our purposes, it is not critical to resolve this discussion here. Rather, the important point is to illustrate that a socio-technical niche or regime incorporates many dimensions of change and, if we collect the lowest-level elements given in the tree diagram in Figure 3.3 (i.e. those elements on the right), we have a list of these dimensions. Using these, we can analyse, systematically, socio-technical trajectory changes (see Figure 3.4). And, in principle, we could use these to map the complete evolution of any socio-technical trajectory.

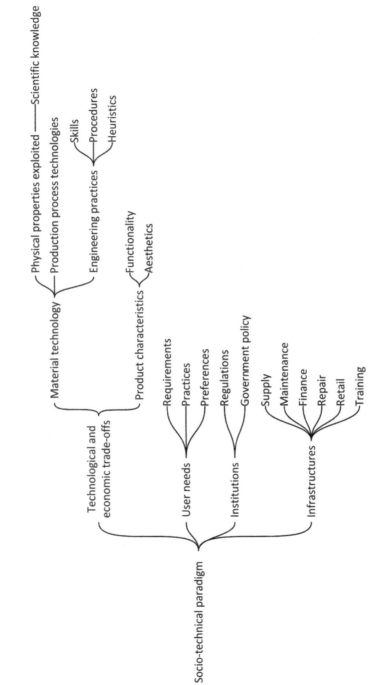

FIGURE 3.3 A socio-technical trajectory and its dimensions

Source: Redrawn from Byrne (2011, p. 36).

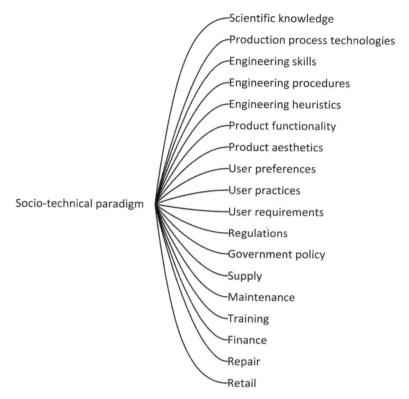

FIGURE 3.4 The dimensions of a socio-technical trajectory
Source: Redrawn from Byrne (2011, p. 37).

Regarding how we recognise changes in trajectory, we can use our theorising of expectations, visions, and first- and second-order learning, discussed above. There, we argued that second-order learning leads to a change of expectations, and this we equated with a change of direction of the trajectory; first-order learning envisions expectations giving 'movement' along that trajectory dimension. This is shown schematically in Figure 3.5.

Thus, we may initially be on a trajectory along which we are envisioning 'Expectation 1' through first-order learning, where the expectation might be that SHSs could be rural electrification solutions and first-order learning would include what kind of functionality users prefer their systems to provide. At some point, we experience second-order learning, and this stimulates a new expectation. Here, we might realise in our field trials that SHSs might not suit the needs of certain people, such as the Maasai, who follow a semi-nomadic lifestyle, and so solar lanterns might be more appropriate. This 'generates' a new expectation – 'Expectation 2' – that now acts as the guide for the direction of learning, rather than our initial expectation. This is not to say that the first expectation (for SHSs) will have been dropped.

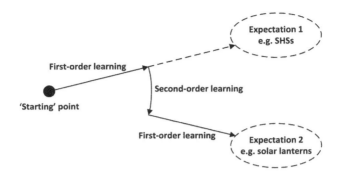

FIGURE 3.5 First- and second-order learning with expectations
Source: Adapted from Byrne (2011, p. 37).

It is quite possible that others will continue to pursue it, while the alternative expectation (for solar lanterns) is the focus of parallel learning effort.

Now, with an elaborated notion of a socio-technical trajectory in mind, we can clarify what we mean by systemic intervention, and also begin to see why the current focus of much scholarship and policy on hardware and finance is inadequate. Beginning with the latter point, analysing only technologies and finance neglects large parts of a socio-technical trajectory and leads to policy interventions that are likely to be piecemeal. This might be sufficient when we are considering regime dynamics, as it might be argued that the rest of the socio-technical trajectory can be taken for granted. User practices, it might be argued, do not need attention, as they can be assumed to be unchanging. This might be a reasonable assumption, although it would miss identifying potential windows of opportunity to effect more profound change through policy interventions. However, in the case of socio-technical niches, where effort is needed to build the substance of all the socio-technical trajectory dimensions, this partial view becomes problematic. If we accept the argument that each of these dimensions co-evolves with the others, then effective policy interventions will need to act on several dimensions at once. This is what we mean by systemic intervention. The more dimensions that are acted upon at once, the more systemic the intervention. This sets up our final discussion, in which we can bring together these various concepts into our notion of socio-technical innovation systems.

Socio-technical innovation systems

In this section, we draw together the insights from this and Chapter 2 to articulate how an Innovation Systems perspective can be broadened to account for insights from the Socio-Technical Transitions literature. This provides us with an over-arching conceptual framework whereby *socio-technical innovation systems* represent the contexts within which transformations in sustainable energy access (or indeed broader transformations based on sustainable innovation and socio-technical change) can be nurtured. Socio-technical innovation systems provide the context where relevant knowledge flows and the development of technological capabilities are nurtured, and where niches of sustainable energy technology adoption can be fostered to the point where they begin to compete with fossil fuel-dominated unsustainable energy regimes. We expand the limited definition from the Innovation Studies literature given towards the beginning of Chapter 1 to provide a broad definition of what constitutes a socio-technical innovation system, synthesising relevant insights from both the Innovation Studies and Socio-Technical Transitions literatures discussed in this and the previous chapters.

As we saw in the discussion in Chapter 2, the majority of innovations in any context tend to be of an incremental form, where continuing efficiency gains are made over time, or of an adaptive form, where existing technologies are adapted to work in new countries, industries, firms, farms or households. For example, incremental efficiency improvements characterised the Korean steel industry, which eventually moved to the international technology frontier (D'Costa 1998; Gallagher 2006), while adaptive innovation of the internal combustion engine was what facilitated Brazil's international leading role in transport-related biofuels (Lehtonen 2011). This could equally apply, for example, in the context of a farmer in Sudan adopting water-efficient farming techniques and adapting them to their specific environmental conditions, or an entrepreneur in Kenya configuring small waste solar panel parts to create a business in supplying solar PV modules to charge mobile phones (as we discuss in Chapter 4).

Of course, all countries aspire to possess the capabilities to create more radical innovations in the hope that these will bring higher value-added economic returns. And, certainly, there are countries that have achieved these goals in recent history, despite having begun their development pathways from a position of extreme poverty. However, the pathways these countries built began with laying foundational capabilities first – incremental innovation capabilities in basic engineering and managerial competence. Alongside these foundational capabilities, they focussed on building and strengthening their innovation systems. As they did so, they were able to absorb more complex technologies and begin to further develop these indigenously. In time, this meant more attention to higher technology R&D, often in collaboration with international firms, and subsequent positions at the 'frontiers' of certain technologies. Hansen and Ockwell (2014), for instance, demonstrate such learning processes, and the role of interactions with foreign technology owners, in the context of the Malaysian biomass power equipment industry. In contrast, those countries that attempted to develop their R&D capabilities before laying any foundations were generally unsuccessful, delaying their 'catching up'.

For most developing countries – and certainly for the least developed – only weak or highly fragmented innovation systems currently exist. Consequently, while they may be implementing some projects through international policy instruments, such as the Global Environment Facility (GEF), they find it difficult or impossible to attract the kind of projects available through market-based instruments, such as the CDM. Even where some of them are now implementing small CDM projects, it is not clear whether they are able to leverage any further development benefits. For example, Kenya has attracted some CDM projects to replace incandescent lamps with the more energy-efficient compact fluorescents, but there appear to have been no attempts to move beyond the simple adoption of these to some kinds of assembly activities that might enable the capture of higher value-added capabilities (Byrne 2013a). This perhaps reflects weak policy-related capabilities, as well as a lack of any appropriate existing industrial activities that might be exploited to move into such production. Whatever the case, it may represent a missed opportunity, an opportunity that a more systemic approach to supporting such technological change would better support.

While the traditional Innovation Studies literature demonstrates the importance of innovation systems in achieving technology-related economic development goals, the Socio-Technical Transitions literature shows us that we need to be cognisant of more than the skills of, and relationships between, industrial actors. This is particularly so when we are interested in creating entirely new technological or innovation trajectories that profoundly challenge the power of established technological systems (or socio-technical regimes). It is from this perspective that we can understand that cultures of practice are as much a part of the inertia of these established systems as the hard technologies and the actors who benefit from them. For example, attempts to address any of the problems associated with the use of cars for personal mobility, such as raising the price of fuel or city-centre parking fees, are met with fierce resistance from consumers as much as from producers. In other words, the interconnectedness and interdependencies that are needed for a well-functioning innovation system can also facilitate inertia in that system. Geels (2002, p. 1260) gives an indication of the interconnections that constitute such a system (in his case, a socio-technical regime: see Figure 3.6). But this

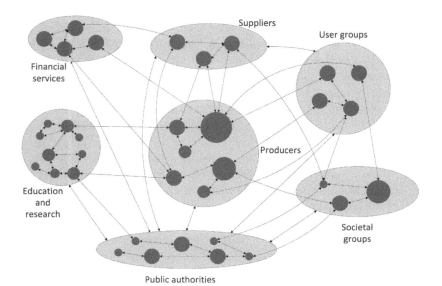

FIGURE 3.6 The multi-actor network involved in socio-technical regimes
Source: Adapted from Geels (2002, p. 1260).

diagram could just as easily represent certain aspects of a 'good' socio-technical innovation system.

This points to the need to develop socio-technical innovation systems carefully, so that they are more likely to be sustainable, particularly if we are interested in development pathways around newer, sustainable energy technologies for which there are many uncertainties. Niche theory, as discussed above, offers particular hope here. Although niche analyses often concern themselves with radically new or highly novel innovations, niche activities consist primarily of incremental learning in real-world settings. As such, there is the chance that all the groups represented in Geels' (2002) socio-technical configuration, if brought together for 'problem-solving' in particular contexts, can develop workable 'solutions' – or innovations – in a co-evolutionary process that meets their different needs.

We can therefore build on the four sets of generic activities – or processes – that niche theory identifies (described in detail above) as necessary to pursue in order to develop such workable solutions or innovations. These can usefully be grouped into the following four categories:

1. learning and creating knowledge
2. institutionalising learned socio-technical practices
3. growing and developing constituencies of support
4. consensus-building.

In many ways, these activities fit in well with what we know about innovation systems from the Innovation Studies literature discussed in Chapter 2 and what we so far understand about how to nurture their development. *Learning and creating knowledge* about particular innovations most clearly relates to the building of cap-abilities and is crucial for the continued evolution and development of an innovation system that is creative and adaptive. Such learning needs to be fostered around the many dimensions of an innovation (the multiple 'socio-technical fronts' that constitute a socio-technical trajectory) – not just the technical and economic, but also in terms of relevant user preferences and practices, government policies, politics and the infrastructure in which the innovation is to be used. *Institutionalising learned socio-technical practices* is the process by which the capabilities developed during learning are widely adopted. It also refers to the development and instigation of relevant policies, laws and regulations, as well as less formal 'rules', such as cultural norms. *Growing and developing constituencies of support* refers to actively encouraging the growth of networks of actors who invest various resources (financial, knowledge, political, etc.) to help realise the success of the innovation. These actor networks can form the basis upon which the linkages and feedbacks necessary for a detailed under-standing of societal and market needs can be built, as well as enhancing the more general flow of knowledge resulting from learning. They also help in *consensus-building* by providing the channels through which expectations about the innovation and its role in 'solving' particular societal needs can be discussed, negotiated, contested, and so on. Consensus-building can also justify certain kinds of practice over others, providing – through institutionalisation processes – the cultural norms to which firms can respond when developing specific innovations. It can also perform more overtly political work, pressuring policy-makers into designing institutional frameworks that support specific innovations. Of course, there are no guarantees of a smooth ride in these processes – and some innovations will inevitably fail – but their incremental nature does at least leave space for adaptation to changing circumstances (whether these are social, technical, political, environmental or otherwise).

Conclusion: towards pro-poor socio-technical innovation system building

To summarise, we can define in broad terms what a well-functioning socio-technical innovation system looks like – one which maximises opportunities for nurturing transformative change around sustainable energy (or other) technologies. Based on the literature discussed in Chapter 2, an innovation system is made up of 'inter-connected firms, (research) organisations and users all operating within [a national] institutional environment that supports the building and strengthening of skills, knowledge and experience, and further enhances the interconnectedness of such players' (Byrne *et al.* 2012a, p. 1). But we can extend this definition by considering the socio-technical nature of innovation and technological change. A focus on socio-technical innovation systems (as opposed to simply innovation systems) allows the adoption of the most promising insights from both the Innovation Studies-inspired and Socio-Technical Transitions literatures described in this and Chapter 2. It goes beyond the limits of the Innovation Studies literature by focussing analytical and policy attention on several key aspects of innovation and technological change. These include attending explicitly to: (1) the role of technology users; (2) the co-evolutionary nature of technological change, innovation and social practice; (3) the extent to which sustainable energy technologies fit with or stretch existing social practices around energy use; (4) the competition sustainable energy niches face from socio-technical regimes; and (5) the analytic and normative considerations highlighted in the niche literature as key issues through which sustainable energy technology adoption by poor people in developing countries might be analysed and actively pursued. We can also add to this the international dimension. That is, any well-functioning national socio-technical innovation system will also be connected internationally through market, social and political relationships and, indeed, these are essential for the continued flow and development of knowledge, skills and innovations.

As such, we arrive at a more comprehensive definition of the systemic context within which the kinds of transformation in sustainable energy access implied in policy ambitions such as SE4All might be realised. This suggests the hypothesis that:

> Hypothesis 1: Transformative change in the adoption of sustainable energy (and other) technologies happens as a result of the development of well-functioning socio-technical innovation systems.

This can be extended to further hypothesise that:

> Hypothesis 2: Where project-by-project or hardware financing interventions, or private sector entrepreneurship financing approaches (as characterised in Chapter 1) are observed in practice, they will have met with limited success.

Such interventions will not be able to explain examples where transformative changes in sustainable energy access are observed. Note that we are not saying that finance, or functioning technologies, or private sector entrepreneurs, or projects that seek to implement or support these are unimportant. Rather, we wish to stress that they are insufficient to drive transformative change in and of themselves, and will only work where they play a linked and coordinated role in broader processes of Socio-Technical Innovation System Building around sustainable energy technologies. Indeed, as the detailed historical account of the development of the off-grid solar PV market in Kenya in subsequent chapters demonstrates, such Socio-Technical Innovation System Building activities often precede the entrance of the private sector, providing the bedrock upon which markets can subsequently grow and in the presence of which hardware financing and private sector entrepreneurship initiatives are more likely to stimulate market growth. These hypotheses also allow for empirical work that can test the limits of the literatures reviewed here and in Chapter 2 – specifically in the contexts of low-income countries and energy access for poor people – extending and perhaps radically revising them as a result. Concerns regarding space and spatio-cultural contingencies, including geography-inspired critiques of Transitions scholarship (e.g. Lawhon and Murphy 2012), are thus explicitly introduced into the analysis.

As we have emphasised at several points so far in this book, however, policy ambitions such as SE4All also imply that transformations can be driven by deliberate interventions through policy and practice. In this sense, the efforts of different actors who play a role in nurturing socio-technical innovation systems become of interest, leading to an additional hypothesis that:

> Hypothesis 3: Transformative change can be, and has been, driven by the efforts of key actors through policy and practice.

As indicated above, this might be theorised along similar lines to Deuten's (2003) cosmopolitan actors, although, as argued above, we refer in this book to such actors as *socio-technical innovation system builders*. This additional proposition is particularly powerful in its implications for policy and practice. Essentially, if we can better understand the nature of socio-technical innovation system builders, their actions and the contexts in which these actions work, then policy interventions could be designed to play exactly this kind of Socio-Technical Innovation System Building role. In other words, policy and practice in developing countries could seek to replicate the role and actions of socio-technical innovation system builders by implementing better-coordinated, deliberate systemic interventions. Given the urgency of the social problems of poverty and climate change, and the accompanying policy ambitions that provide the context for this book, the potential impact of this is significant.

It is important to flag here, however, the discussion in Chapter 7 of this book regarding the need to draw together various ideas that relate to the notion of socio-technical innovation system builders – ideas being discussed in disparate ways in both the Innovation Studies and Socio-Technical Transitions literatures. These include actors who connect together different parts of an innovation system, referred to as 'systemic intermediaries' (van Lente *et al.* 2003; Kivimaa 2014); actors whose efforts play a role in driving cumulative causation (often borrowing from Political Science ideas such as 'advocacy coalitions' and 'policy entrepreneurs') (Kern 2011); 'technology advocates', who do socio-political work to empower socio-technical niches, including by constructing actor networks (Smith and Raven 2012); as well as the cosmopolitan actors mentioned here who do socio-cognitive work to render technologies more widely applicable (Deuten 2003; Geels and Deuten 2006). We therefore suggest that a key research need is to conduct significant further work to connect and develop these threads into a comprehensive theoretical perspective that more fully captures the nature and role of such actors, including the often underplayed political nature of their efforts. Moreover, this theory needs to be applicable in the context of concerns over transformations in the use of sustainable technologies (energy or otherwise), as well as in the contexts of the many poor developing countries. So far, much of the theorising has been done in relation to industrialised countries and rapidly emerging economies. For those countries that wish to build – or, in some cases, create – innovation systems, the theorising will need to go well beyond understanding how actors connect together bits of existing systems. It will not only have to be situated within a systemic perspective. It will also need to understand the much more active and creative role that such intermediaries, advocates or entrepreneurs can and do play, and how transformations derive from such actions. But, pushing further, the theory will also need to engage with the governance challenges that inevitably arise if a small number of key actors are to be tasked with building socio-technical innovation systems. We hope that this book constitutes a first step in this direction and can inspire future research and practice that works to develop this further through grounded historical research, as well as through action research engaged directly with policy and practice.

Finally, we need to reassert the positioning of these theoretical propositions within the context of the Pathways Approach, which, as described in Chapter 1, provides the overarching heuristic and our normative positioning in this book. It should – we hope – be clear from Chapter 2 and Chapter 3 that the theoretical positioning of our argument represents a fundamental departure from both the dominant two-dimensional engineering-and-economics focus of most existing academic literature on energy access in Sub-Saharan Africa and other development contexts. It also implies a fundamental departure from the framing of the majority of existing international sustainable energy access policy, namely, the dual dominance of Hardware-Financing market-based mechanisms and financing for Private Sector Entrepreneurship. The reframing of the problem of sustainable energy access as one requiring the building of socio-technical innovation systems implies an approach involving multiple actors, extending well beyond the firm and industry focus of the traditional Innovation Studies literature and the majority of policy discourses around technology and development. Perhaps most important, this extension must include the poor people, who, it is assumed, will benefit from access to low carbon technologies, as well as the social practices and associated processes of human and economic development that such technologies are assumed to support. This points to a more plural, dynamic and political perspective on how transformations in sustainable energy access might be facilitated and places centre-stage a concern with the lived experiences of poor people and the socio-cultural contexts within which such transformations will unfold. But it simultaneously deals – especially through the use of innovation systems thinking – with broader economic development concerns. As such, it is arguably capable of offering ways forward with the potential to deliver against multiple needs, from those of basic human development and social justice to aggregate economic growth and indus-trialisation concerns. Of course, it must be emphasised that such potential carries no guarantees, remaining contingent on how processes of governance, power and politics play out in practice. Indeed, at its heart, such potential requires considera-tion of the role of the state in development and of concerns that may be perceived to be in tension with the neo-liberal code for how development is – or should be – 'done' (as convincingly portrayed by Selwyn 2014).

We return in Chapter 7 to this discussion of pathways and the governance impli-cations of both the theoretical propositions and the empirical analysis in this book. In the intervening chapters, we explore some of the insights that can be realised by applying many of the concepts we have so far discussed to examples of sustainable energy access in practice. In particular, we provide an in-depth historical analysis of the oft-lauded success of the off-grid solar PV market in Kenya. Through this, we are able to explore the extent to which the different framings identified so far on how to foster sustainable energy access have played out in practice in what is currently one of the few examples of something approaching 'transformative' change in relation to sustainable energy technology adoption in a low-income country context.

Notes

1 The use of the word 'need' here is not meant to imply some fundamental human need, although that may be a reasonable interpretation for some socio-technical regimes. Rather, it refers to the whole range of wants, needs and demands within the relevant social context.
2 In fact, the socio-technical view would consider supply and demand together as a co-construction or co-evolution (Geels 2004, p. 898).
3 Non-users may be those who would be affected by a technology even though they are not directly participating in its use, an example being householders near a wind farm (Raven 2005, p. 40).

4

EMERGENCE AND ARTICULATION OF THE KENYAN SOLAR PV MARKET

Introduction

Having established the normative and theoretical basis of the book, in subsequent chapters we explore these various claims through an in-depth historical account of the development of the off-grid solar PV market in Kenya. This history is presented in two parts. In this chapter, we deal with the emergence of the market through to its maturation as the largest per capita market for off-grid PV in the world (Ondraczek 2013). In Chapter 5, the second part of the history, we start with an analysis of solar PV niche interactions with the policy regime in Kenya and end with an account of contemporary advances in the market, including the newly established market for pico-solar lanterns and the establishment of Kenya's first-ever solar PV assembly plant.

Throughout the analysis, we use the concepts from Strategic Niche Management (SNM) described in Chapter 3, with socio-technical analysis following each broad segment of the history we present. The theoretical concepts concerning socio-technical innovation systems and the implications of the empirical analysis here in relation to existing framings of the sustainable energy access problematic, as well as broader concerns with low carbon development, green growth, and so on, are then taken up in the analysis in Chapter 7.

All of the data analysed here and in Chapter 5 were collected using a combination of literature review (both grey and published) and semi-structured interviews and stakeholder workshops in Kenya, conducted in two periods: the first between 2007 and 2008, and the second between 2012 and 2014. This was developed using a variation on Douthwaite and Ashby's (2005) innovation histories method, which is a participatory approach for engaging with stakeholders to develop a detailed historical account (the 'innovation history') of how certain innovations in the use of technologies emerge within specific contexts. In our case, we were specifically interested in the emergence of the off-grid solar PV market in Kenya. For more methodological details, see Byrne (2011) and Byrne *et al.* (2014b).

Why look at solar PV in Kenya?

As emphasised in Chapter 1, the ambitions of initiatives such as SE4All of universal energy access by 2030 imply nothing short of a transformation in levels of energy access, particularly in Sub-Saharan Africa, where access levels are as low as 20 per cent in many areas. Certainly, very few energy access examples exist in Sub-Saharan Africa that could be described as such a transformation. The success of the off-grid solar PV market, however, is widely hailed as being one such transformative example. This pertains mostly to solar home systems (SHSs: solar PV panels charging batteries to run a number of electrical appliances such as lights, mobile phone chargers, radio, TV) but in recent times solar portable lanterns (SPLs or pico-solar lamps: lamps with accompanying PV panels for recharging, and sometimes supporting mobile phone chargers) have also become significant and are therefore included in this analysis.

The market for off-grid solar electricity in Kenya, especially SHSs, is one of the most significant in the world. In terms of both annual sales and overall numbers of SHSs installed, Kenya accounts for around 10 per cent of the global market, making it second only to China in aggregate terms, and number one in the world on a per capita basis (Ondraczek 2013). There are estimated to be well in excess of 300,000 SHSs in Kenya sold through a vibrant private market. Explaining this widely lauded success becomes a particularly germane focus for attempts to try to learn lessons in how driving transformations in sustainable energy access elsewhere might be attempted. Moreover, doing this in a way that applies and tests the theoretical framework developed in the last two chapters is of particular importance in terms of attempting to generalise from this case to inform research, and to potentially guide policy and practice in different contexts.

Debunking the myth of an unsubsidised market

The dominant narrative in regard to the Kenyan solar PV market – repeated in both the grey literature and in academic publications (e.g. see van der Plas and Hankins 1998; Jacobson 2007; Ondraczek 2013), and echoed by many donors and international observers – is that it is an example of private sector-led development and often is described as an unsubsidised market. This is perhaps not surprising, given Kenya's vibrant private solar PV market and the resonance of this narrative with neo-liberal development ideals (on the latter, see Selwyn 2014). Also not surprising is the support that such a narrative provides for the kinds of Hardware-Financing and Private Sector Entrepreneurship framings critiqued in Chapter 1. Indeed, as we shall see in this chapter and Chapter 5, such approaches have been and are being applied in Kenya (e.g. via the Photovoltaic Market Transformation Initiative and the more recent establishment of a Climate Innovation Centre).

However, as our detailed analysis demonstrates, an alternative interpretation of this market's development fundamentally challenges both the 'private sector-led development' narrative and the 'unsubsidised market' description. According to this alternative reading, the development of the market looks like something much closer to the Socio-Technical Innovation System Building described in the last two chapters. This so-called 'unsubsidised market' has, in fact, relied heavily on donor support to enable learning and capability-building in the PV niche. In other words, the private sector-led development narrative has, ironically, been a powerful tool to attract resources from donors to subsidise development that has supported the activities of private sector actors. Without these subsidies, it is highly unlikely that the long-term learning and capability-building that actually underpinned the creation of the market would have happened, and it is therefore debatable whether there would ever have been a Kenyan solar market phenomenon.

But the market for SHSs and for pico-solar lamps has indeed grown, and this has recently attracted investment in a solar module assembly plant, the first in Kenya. It is too soon to know whether this plant will be a success, but it does raise interesting questions and speculation about the directions of development the PV pathways will take in Kenya, intersecting in important ways with debates on how policy interventions can work to drive both access to sustainable energy technologies as well as providing conditions under which industrial and aggregate economic growth concerns can be addressed. We return to these and other issues emerging from the analysis, especially in relation to our theoretical concern with Socio-Technical Innovation System Building, in Chapter 7.

What of the decreasing price of solar PV?

One final point that it is important to tackle head-on before embarking on this analysis is the issue of the deeply declining price of solar PV in recent times. A total decline in PV module price of 50 per cent occurred in the five years leading up to 2015 (Chattopadhyay *et al.* 2015). Notwithstanding some poor quality modules on the market, there has also been a general increase in the availability of more efficient end use technologies, such as light-emitting diodes (LEDs) and direct current (DC) appliances that increase the utility of PV systems and their efficiency in terms of running appliances. It would be disingenuous to ignore the impact that the improved economics and utility of solar PV will have played in increasing the size of the markets across the world, including in Kenya.

This price reduction, however, has been observed internationally, and yet certain countries have done significantly better than others in terms of creating and growing markets for solar PV. This international differentiation in the diffusion of solar PV raises important questions as to why some countries' markets have grown so quickly and extensively while others have not. For example, why is it that Kenya's market has grown so much more quickly and become so much larger than that of neighbouring countries like Uganda and Tanzania? In 2011, Tanzania was estimated to have an installed solar capacity of 5MWp (Meza 2013) and Uganda, in 2009, was estimated to have around 1.1 MWp (Hankins *et al.* 2009). In comparison, Kenya's installed solar PV capacity is in the region of 16 MWp (Meza 2013) to 20 MWp (Hansen *et al.* 2015). Clearly, all things cannot be assumed to be equal, and the falling price and increasing utility of solar PV globally cannot be assumed to explain, in and of themselves, the remarkable story of the growth of the off-grid solar PV market in Kenya. Moreover, the Kenyan market was already doing well long before the recent declines in PV prices. For these reasons, detailed historical analyses that take a more systemic perspective (of the kind we attempt in this book) potentially have much to contribute in informing policies and practices aimed at transformative interventions.

How the next two chapters proceed

In this chapter, we begin by charting the arrival of PV systems into East Africa through donor-funded community services projects that helped create seeds for the later emergence of a household market in Kenya. Following this, we discuss how the household market potential began to be exploited and how the idea was picked up by other companies. By the time Mark Hankins (who, as will become clear from the analysis below, became an important actor in the Kenyan PV niche) did his MSc research into the Kenyan PV market in the late 1980s, there was already significant activity in household PV systems. Hankins set about dis-seminating this and recruiting others to create a broadening network of actors. Eventually, he had the chance to do more substantial projects and started his own company – Energy Alternatives Africa – to exploit the opportunities. The chapter goes on to describe and analyse these and other niche developments that took place as the market grew in Kenya.

Eventually, in the early part of the 2000s, niche actors were interacting directly with the policy regime as they attempted to influence Kenya's new energy policy. This had mixed results and reflects an uneasy relationship between niche actors and some influential figures in the policy regime – an uneasy relationship that in some ways continues. More recently, a market for pico-solar has emerged in parallel with that for SHSs, drawing in new actors and offering new hope of addressing the needs of poorer groups. Kenya has also attracted investment in its first-ever solar module assembly plant. These latter developments are dealt with in the second part of the innovation history of solar PV in Kenya, presented in Chapter 5.

PV for community and commercial services

Let us begin, then, by looking back at the early days of solar PV in Kenya. During the late 1970s or early 1980s, it seems that PV was already in use in Kenya for some limited applications, although it is difficult to establish exactly what these applications were, and when and by whom the systems were installed. According to Hankins and Bess (1994, p. 2) and Duke *et al.* (2002, p. 481), these early systems were used to power telecommunications. It is not clear whether they were commercial, funded by donors, or whether the Kenyan Government was involved in some way, and the Kenyan National Paper for the 1981 UN Conference on New and Renewable Sources of Energy (held in Nairobi) simply states that PV had 'barely been tried in Kenya' (Mugalo 1981, p. 10). Whatever the precise details, it seems that supply and installation in the Kenyan PV sector – to the extent that they existed – were dominated by international telecommunications companies and that 'all the PV components used including the wiring accessories were imported' (Masakhwe 1993, p. 66).

The initial experiments with systems in Kenya, for which specific information is available, were for clinic and vaccine refrigerator systems (McNelis *et al.* 1988; Roberts and Ratajczak 1989). The first two of these systems (in clinics) were installed in Kenya in May 1983, one each in the villages of Ikutha and Kibwezi (Roberts and Ratajczak 1989, p. 15, Table II). In September 1984 and January 1985, three vaccine refrigerator systems were installed (McNelis *et al.* 1988, p. 43, Table 4.3), although no locations are given. The two clinic systems were funded by USAID, used equipment manufactured by Solarex (a US company), and were designed and installed by staff from the NASA Lewis Research Center (NASA-LeRC) (Roberts and Ratajczak 1989). The three vaccine refrigerator systems were funded through the World Health Organization Expanded Programme on Immunization (WHO-EPI) effort and used BP Solar-LEC equipment (McNelis et al. 1988). The objective of the USAID clinic project was (Roberts and Ratajczak 1989, p. 16, Table III):

> to increase health services in rural areas by demonstrating applicability of PV power systems for rural clinics by providing electricity for vaccine storage, lighting and other discretionary uses; e.g., dental equipment, communications, staff residential lighting and water pumping.

These projects were part of much larger programmes of experimentation with PV systems in developing countries. During the period January 1983–October 1984, systems of various kinds were installed by NASA-LeRC in nine Sub-Saharan African countries: Burkina Faso, Gabon, The Gambia, Ivory Coast, Kenya, Liberia, Mali, Zaire and Zimbabwe (Roberts and Ratajczak 1989, p. 15, Table II). The types of systems installed were: rural clinics, vaccine refrigerators, school lighting and TV/VCR, water pumping and outdoor lighting (Roberts and Ratajczak 1989, p. 14, Table I). A further six vaccine refrigerators in total were installed in three countries (Ghana, Kenya and Tanzania) through WHO-EPI between May 1984 and January 1985 (McNelis *et al.* 1988, p. 43, Table 4.3). And, prior to these, in 1980, Oxfam had supplied 52 PV-powered pumping systems to Somali refugee camps (Hankins and Bess 1994, p. 2). The WHO Expanded Programme on Immunization had the highly ambitious goal of immunising[1] 'all children of the world by 1990' (Henderson 1989, p. 46). Part of this effort was to strengthen the cold chain, hence the interest in PV-powered vaccine refrigerators; field tests suggested they were more reliable than kerosene-fuelled types and were cheaper under certain conditions (McNelis *et al.* 1988, pp. 45–47). By the mid-1980s, the programme included a commitment 'to adopt PV-powered refrigerators wherever they were economically and technically justified' (Foley 1995, p. 12).

Perhaps as a direct response to this donor-funded activity, and in anticipation of the future PV markets in developing countries, a few companies set up offices or agents in Kenya (and elsewhere) during the early 1980s. Certainly, some of the market projections at the time would excite significant business interest. For example, a study for the European Commission reported estimates of market *potential* of different PV applications in developing countries at 50 million small solar water pumps, 5000 vaccine refrigerators per year, and a total village power potential of 3.75 TWp (Starr and Palz 1983, p, 121). Animatics, an agricultural equipment supplier in Kenya, started supplying ARCO modules as early as 1981, selling a reported 420 modules that year (Hankins 1990, p. 62, Table 4.1). BP Solar set up an office in Nairobi, or possibly established Securicor as their agent, in 1983 (EAA 1998, p. 23). Total Solar entered the market in late 1985 (Hankins 1990, p. 67), although they may have been concentrating on establishing a network of dealers for the first year (Rioba 2008, interview); indeed, Energy Alternatives Africa (EAA) (1998, p. 25) state that Total became 'active' in 1987. Telesales, a retailer, may have entered the market in 1985 as well (EAA 1998, p. 24), although Abdulla (2008, interview) claims that they had been stocking modules since the late 1970s. And Alpa Nguvu entered the market around 1986 (Hankins 1990, p. 60).

It is unclear the extent to which the Kenyan ministry responsible for energy was aware of these various developments, although Kenya did have a Ministry of Energy (MOE) by 1979 (Goodman 1984). We know that, around the mid-1980s, the Italian government donated some PV equipment to Kenya, including a water pumping system and energy research laboratory (Rioba 2008, interview). The MOE created a biomass department around 1984 (Arungu-Olende 2008, interview), which eventually became a renewable energy department in 1998. The first Kenyan Energy Policy was written in 1987, but it remained an internal document (ROK 1987). While the MOE was engaged in renewable energy projects, and the 1987 policy at least mentioned PV, there is no evidence that it was particularly active with the technology (Rioba 2008, interview). For the most part, the MOE and donors were more concerned with finding solutions to the problems with biomass energy.

So, in the early 1980s, there was no significant market in Kenya for household PV systems. Projects for commercial and community services systems continued, and so a market developed around these – a market that continues to account for a large part of the installed capacity of PV systems in the region (ESD 2003).

An emergent market trajectory for household PV

This section describes and analyses the very early period of the household market in Kenya: how the market emerged and how it was initially developed. These activities attracted others and, as we will see in subsequent sections, the initial work had a lasting effect on the PV niche in Kenya.

Discovery of a household market

The private market in household PV systems is said to have started during 1984 and its beginning is attributed to the activities of Harold Burris, an ex-Peace Corps volunteer, after he set up the company Solar Shamba[2] in a coffee-growing region south of Mount Kenya (Acker and Kammen 1996, p. 87; Duke *et al.* 2002, p. 481). Burris was an engineer by profession and, according to one obituary, had worked in the nascent US solar industry (SolarNet 2001, p. 8), particularly with Texas Instruments (Hankins 2007, interview), before coming to Kenya with the Peace Corps in 1977 (Perlin 1999, p. 132). He was, according to Hankins (2007, interview), politically radical and fiercely independent, and so found it difficult to work within the constraints of traditional organisational hierarchies. As a result, he tended not to keep a job for very long before either resigning or being dismissed; indeed, according to Hankins, he was dismissed from his Peace Corps assignment, following which he returned to the US, where he did some 'early computer-circuit work and helped to develop a health device for a friend', which made him enough money to return to Kenya, around 1979, with his own resources (Hankins 2007, interview). After spending some time in Mombasa with his wife, Stella, the couple moved to Yata (Stella's home town) in Machakos where Burris began working with appropriate technologies (ATs) and became 'well-connected with AT people' there (Hankins 2007, interview). He attended the UN Conference on New and Renewable Sources of Energy in 1981, where it is likely, according to Hankins, that he used the opportunity to network extensively. In 1982, Burris set up *Kidogo*[3] *Systems* (Jacobson 2004, p. 125, n. 160) and tried, unsuccessfully, to market a PV-powered sewing machine through the Singer Company (Hankins 1993, p. 31; Perlin 1999, p. 133). He had developed the idea for his wife, a seamstress, powering her sewing machine by PV 'from day one' (Hankins 2007, interview). While Hankins (1993, p. 31) seems to attribute the failure of this project to an abortive coup in Kenya in 1982, he now says it failed because the machine was far too expensive for the Kenyan market (Hankins 2007, interview). Nevertheless, the episode indicates that Burris was searching for a means to earn a living in Kenya, that he was able to source PV equipment, and that he was experimenting with PV systems.

Some time around the middle of 1983, Burris met Mark Hankins by chance at a café in Nairobi (Hankins 2007, interview). Hankins was a Peace Corps volunteer teaching science at Karamugi Harambee[4] Secondary School, which was in the process of considering electrification with a 'used 5 kVA diesel generator' (Hankins 1993, p. 31). The generator was chosen because the cost of connecting to the grid, some four miles away, would have been about USD 21,000 (Perlin 1999, p. 133). When Hankins mentioned this in their conversation, Burris suggested that he could install a PV system instead. Hankins was unconvinced – 'I didn't trust Harry at all; the guy didn't look serious' (Hankins 2007, interview) – but he nevertheless put together a comparative cost analysis for the Board of Governors showing that PV would be cheaper than the diesel generator (Perlin 1999, p. 133). The board was also unconvinced but agreed to visit Burris' home system, after which they were impressed enough to postpone purchase of the diesel generator and to trial the use of PV in four classrooms and the headmaster's office (Hankins 1993, pp. 31–32; Perlin 1999, p. 133; Hankins 2007, interview). The lighting system Burris designed cost the school USD 2000, which 'was less than the first cost of [the second-hand] generator' (Hankins 1993, p. 32). Hankins (2007, interview) recalls that Burris was struggling financially at this time, that 'he was very desperate, he was broke'; the modules he used for the Karamugi installation were left over from his failed sewing machine project. Once the school had given the go-ahead for the installation, Burris went to work on the balance-of-system (BOS) components: charge regulator,[5] 24 VDC lights from a local manufacturer, local car batteries, module mount, and battery boxes (Hankins 2007, interview). Hankins (2003, p. 2) elaborates on which of the BOS components Burris put together himself and which he sourced locally:

> He [Burris] found that ballasts for 12VDC lamps were being manufactured for local buses by a Nairobi company, Sound Communications. Further, he designed and assembled basic charge regulators and DC-DC converters (which allowed use of radios and cassette players) in his own shop. Further, he coaxed the local battery company to improve the design of their automotive battery to make it more suitable for PV systems. He designed module mounting systems and other balance of system components that could be made cheaply and by cottage industry groups.

During the Karamugi installation, which took place some time during the first to third quarter of 1984 (Hankins 2007, interview), 'Burris used the services of an electrician based in the town near Karamugi and he trained the school's lab technician to monitor and maintain the system' (Hankins 1993, p. 32). The results of this monitoring were 'fed back to the installers' (Kimani and Hankins 1993, p. 93).

According to Hankins (1993, p. 32), and Kimani and Hankins (1993, p. 93), the headmaster, some of the teachers and others in the community bought systems for their own homes 'within six months of the school's installation'. This was a clear signal to both Burris and Hankins that there could be a market for household PV systems. Burris 'saw that there was a lot of business and there was a coffee boom[6] going on too so there was a lot of cash' (Hankins 2007, interview).

A major factor in the demand for electricity is the desire to watch television, and portable DC TVs began to appear on the market in about 1981, with the TV signal being broadcast to more and more rural areas during the 1980s (Jacobson 2004, pp. 150–157, and Figures 24 and 26; Hankins 2007, interview). In response to these developments, Burris moved to Embu where he renamed his business *Solar Shamba* (Jacobson 2004, p. 125, n. 160). Hankins, for his part, was already applying to the Peace Corps by the second quarter of 1984 for an independent placement[7] in which he would work with Burris on a project to install PV systems in three more schools, and include in the package the training of local technicians (Hankins 2007, interview). According to Hankins (1999, p. 6), he and Burris believed the training element would be critical to the growth of PV applications in Kenya, and that rural electricians would need to be able to 'sell, install and maintain PV systems'.

In Embu, as he had done in Meru, Burris powered his home and workshop with PV (Perlin 1999, p. 133): 'He was in town but off-grid. ... a kind of in-town appropriate technology demonstration' (Hankins 2007, interview). He now began 'to get heavily into the marketing' (Hankins 2007, interview). Dickson Muchiri, who worked as a sales technician for Burris from about 1986 until moving to the company Alpa Nguvu in 1987/1988, elaborates on the marketing strategies that Burris had developed by that time (Muchiri 2008, interview):

- Writing proposals for organisations looking to get a PV system funded by a donor. If the proposals were successful, then Burris would most likely get the job.
- Placing some kind of 'working sample' in a strategic location such as a small shop. Customers could see it, know that it was working, ask questions, etc.
- Sometimes advertising in local newsletters (although not really in newspapers). There was one that went around in Embu town, for example.
- Participating in dissemination events organised by aid organisations where he could explain solar PV and its potential applications.
- Attending district shows. One example was a show in Embu in 1986 that Muchiri believes achieved wide publicity for Burris. Indeed, van der Plas and Hankins (1998, p. 301) note that 'agricultural fairs were an important information channel in the late 1980s and early 1990s'.
- Using his technicians to cold-call. He employed six permanent and two casual technicians, and whenever they were installing a system in a house, they were instructed to go around the area looking for potential customers. For example, if they saw someone was building a new house, then that person could be a customer. And Hankins (2003, p. 2) reports that Burris 'encouraged these technicians to seek customers among the high-income households on the southern and eastern sides of Mt Kenya'.

Hankins (2007, interview) adds that Burris produced one-page mimeographs, although he does not describe the content of these. We might reasonably assume that these would, at the very least, explain what PV could power and how to contact Solar Shamba in order to buy a system.

By the third quarter of 1984, the Peace Corps had given approval for Hankins' independent placement, providing he concentrate solely on the solar project with Burris (Hankins 2007, interview). Although Hankins says that he had to convince the Peace Corps to approve the independent placement, it seems this was helped by their visiting the Karamugi installation:

The Karamugi installation was a coup: it involved some Peace Corps leaders coming to the school and talking about how this was a great thing. So there was definitely a sense that this was a great idea and so let's talk to the people in USAID about it.

USAID could be expected to be favourable to the idea, as they had already funded a 'very successful' energy project in Kenya in 1984 – the Kenya Renewable Energy Development Project – which saw the creation of the Kenya Ceramic Jiko, an improved small stove[8] consisting of a metal case with a ceramic lining (Hankins 2007, interview). Still, Hankins says:

> I had to write a proposal and design the training and get Harry to go in on this ... Harry was the guy who was dealing with the American companies so Harry was going to get paid to bring that equipment in ... I had to locate three schools. I did a survey of twelve schools; riding around on a bicycle on the eastern side of Mount Kenya convincing schools to put in 50% of the cost ... I did energy audits of the schools; looked at how much wood they were using and tried to come up with a case.
>
> Harry was intimately involved in the process: we would meet in Nairobi in a cheap hotel and we would work on Harry's World War II typewriter and we would do cut and paste as we designed manuals. We also had to identify twelve solar technicians. One was a relative of Harry's wife, Daniel Kithokoi. We got the equipment, identified the schools and we did the installations one after the other [during 1985 and into 1986] ... When we trained the twelve guys, he [Burris] immediately went to all the twelve guys and said 'Be my agent.'

As mentioned above (Muchiri 2008, interview), Burris did employ some of the technicians: six permanent and two casual. By this time, according to EAA (1998, pp. 24–26), Telesales, Alpa Nguvu Solar Systems, and ABM (Chloride) had all entered the PV market – ABM (Associated Battery Manufacturers) being the 'local battery manufacturer' mentioned earlier that Burris had 'coaxed' into improving their automotive battery (Hankins 2003, p. 2), getting the product on the market in 1985 (Hankins 1990, p. 74; Acker and Kammen 1996, p. 88). At the end of the USAID-supported schools project, Hankins and Burris organised a cocktail party in Nairobi so that the technicians could meet these PV and equipment suppliers, resulting in some of the technicians either being employed immediately (Hankins 1993, p. 32) or striking deals with the companies, independently of Burris (Hankins 2007, interview).

Socio-technical analysis of the discovery of a household market

The Karamugi school project

The evidence suggests that, prior to the Karamugi school installation, Burris had not considered PV systems for households as a viable business opportunity. This was despite his using PV for his own home. However, it is clear that he was considering ways to make use of his knowledge of PV to develop a business and had tried to market at least one product: the PV-powered sewing machine. This had failed because it was too expensive compared with the foot-powered device that was already widely available and in use. Even the process of securing the Karamugi installation was a protracted episode; he had failed to convince Hankins, who in turn had failed to convince the Board of Governors, despite having provided a favourable cost-comparison with the proposed diesel generator. It was only after the Board had seen the system at Burris' home and workshop that they accepted PV as a possibility.

We can interpret this slow acceptance by the Karamugi Board quite straightforwardly. PV was a new technology and so it is unlikely that any of the Board members would have seen it in operation before the visit to Burris' home. The other ways of getting electricity – the grid or diesel generator – were already somewhat familiar. This would have made PV seem highly risky or, at least, unproven. Indeed, they may not have had any conception of PV. Seeing a system in operation would have demonstrated its functionality and may have instilled some confidence that Burris was someone who could perform the installation competently. Certainly, the Governors were now willing enough to take the risk. From an SNM perspective, if second-order learning is characterised by changed assumptions, then we could say that the Governors experienced such learning because they now included PV as a possible source of electrical services, alongside the grid and diesel generators. Whether this was a change of assumptions or not, we can certainly claim that they were able to form a detailed socio-technical vision – a well-articulated cognitive schema of PV-generated electricity services. Moreover, that vision was now grounded in a physical reality that was close to their personal experiences.

Once the system was in use in Karamugi, further learning occurred that we can most likely categorise as first-order. Obviously, there would have been much learning about the operation, maintenance and monitoring of the system – clearly learning of a first-order quality. But there would also have been the issue of confidence in the technology. For some, this confidence grew quickly and was strong enough that they were willing to buy systems for their own homes.

From the point of view of Burris, witnessing the impact his home system had on the decision-making of the Board may have been an important experience that contributed to his later marketing strategies. Despite his having supplied the Board with a quantitative assessment of the costs of a PV system compared with a diesel generator (assuming that he had at least some hand in this, as Hankins would have had to get information about a PV system from someone), the decision to buy a system was not made until the Board had actually seen one in operation. Of course, the visit to Burris' system suggests that the cost-comparison had raised their interest to some extent. But the 'deal-maker' seems to have been the system visit. This deal-making quality of demonstrations was reinforced by the Karamugi installation itself, when the headmaster and others ordered systems for their homes, and other schools became interested. Hankins, if it had not happened already, was also convinced by the Karamugi installation,[9] and was inspired to work with Burris on another project, this time much larger. Further, it was now clear that a household market in PV systems was a realistic possibility.

We can infer from these events that learning of various kinds occurred. Burris was certainly engaged in first-order learning in terms of the technical details of the systems; he had spent some effort putting together the BOS components, and he was receiving information on the performance of the Karamugi system. But we can also infer some second-order learning for Burris regarding the possibility of a household market. He had not tried to market household systems, as far as we know, even though he knew from personal experience that they were technically feasible. One explanation of this is that he assumed there was no market. However, once a demand was demonstrated to him, 'he mobilised very quickly' (Hankins 2007, interview). He already had a well-articulated technical vision of PV, most likely an economic one with a social dimension (in that he used the technology in his own home), and now he was able to add a business or market aspect. He was yet to develop the detail of this market aspect, or how to sell to it, but he had a beginning: there were wealthy enough customers in rural areas who, if they saw the technology in operation, would buy systems for their homes.

Hankins was also recruited to this vision, albeit with his own dimension to it, having now learned that PV was viable and that there was a potential market for PV systems. He could also now see that Burris was 'serious'. Hankins' version of a socio-technical vision included a training aspect. Between them, they had constructed a basic strategy to capitalise on this nascent market: Burris would address the technical aspects while developing his business; Hankins would address the training. This would be more straightforward for Burris in that he could concentrate on finding customers, some of whom were already coming to him. It would have been more problematic for Hankins as he would be unable to sell training in the private sector. So the notion of implementing a donor-funded project that included training would have seemed sensible. Such a project could be expected to replicate the experience of Karamugi: demonstrate the technology to generate interest and then hope customers would emerge.

Hankins was able to recruit relevant people within the Peace Corps to this vision, themselves having been influenced by seeing the system at Karamugi. Again, the demonstration effect was evident. However, the proposed three-schools project was also in line with existing Peace Corps interests. They had been work-ing since 1979, with financial support from USAID, on developing a rural energy survey methodology, which was 'one component of a Renewable Energy Pro-gram ... to assist developing countries in identifying energy needs in rural areas and in implementing alternative, renewable energy projects at the community level' (Peace Corps 1984, p. 7). So, from the Peace Corps perspective, the Karamugi installation was exemplary and it was easy to see that they would support similar projects, assuming some due process, such as a project proposal, and so on. Indeed, the proposed project would be strengthened, in the Peace Corps' view, by a much larger and more systematic training element. This training aspect was also in line with the development regime's interest in building capacity in the private sector.

We can see network-building happening during the Karamugi episode. Burris was already involved in an appropriate technology network in Kenya and knew the PV suppliers, while Hankins was involved in the Peace Corps network and was working in the Karamugi School. Karamugi was deeply embedded in its community, especially considering that it was a Harambee school, and there would have been some connections to other schools, at least, because of the education system. Both the school and the Peace Corps, of course, had access to financial resources: the school directly from the community; the Peace Corps from USAID.

The three-schools project

The processes associated with the three-schools project were sites of further learning, and forming and refining of socio-technical visions; there was also network-building, institutional innovation, and the mobilising of resources. For Hankins, the three-schools project was significant because it resulted in a model of PV market development that he would later use in other countries, as well as much of the material he would use to write what became a textbook of PV system installation tailored to an African context. For Burris, apart from the immediate benefits of paid work and the potential of more to come, the project was important because he was able to train his own agents (as many of them became) at no cost to himself. For the technicians who were trained, the project provided an opportunity to develop new capabilities (skills and knowledge), to get work and to connect with the PV suppliers in Nairobi. The suppliers benefited by gaining access to more trained technicians. The schools, of course, benefited from subsidised PV systems and the electrical services these afforded. And, in terms of a *local* PV niche, the project was important because it demonstrated that PV could be installed by Kenyan technicians – that it did not require highly paid foreign specialists.

We can identify important first-order learning in Hankins' energy audits, which he conducted during his survey of 12 schools. These audits would have helped to quantify aspects of the case he was building to persuade schools to come into the project. The learning here involved developing an energy survey methodology and more precise information on the costs (in time and effort as well as money) of using various energy carriers compared with electricity generated using PV systems. The most direct comparisons would have been with kerosene for lighting and dry cells, fuel-generators or grid connections for electricity. Indeed, Hankins provides cost-comparison examples of all of these, except grid connections, in the 1995 edition of his book *Solar Electric Systems for Africa* (Hankins 1995, pp. 109–112). Not only would these cost-comparisons have been useful in persuading schools to come into the USAID-supported project, they would have helped form the basis for future arguments related to the costs of PV elsewhere, as well as further articulating a PV socio-technical vision.

PV systems were further indigenised through the project. In terms of technical artefacts, there were several innovations. Burris was continuing to refine the technology as much as he could, and he persuaded ABM to modify their auto-motive battery so that it better suited the needs of PV. He worked with others to develop his manually rotatable module mount, which enabled significantly more solar energy to be harvested by a PV system. This was developed with the help of a local NGO. The ballasts for DC lamps were available locally. Burris made his own charge regulators and indicators, and reflectors for the lamps were made locally, as were battery boxes. Clearly, the training of technicians was a significant indigenising process. They were trained in the design, installation, operation and maintenance of systems. Those who worked for Burris would also have been trained in making charge regulators and the other components he developed. And, of course, they would have been active in developing the marketing strategies used by Solar Shamba.

Training, by definition, is about developing practice – an important element of institutional embedding. The design of a system begins with understanding the energy needs of the customer. Here, Burris developed various forms for recording information about a householder's electricity needs, and these were tailored to the kinds of homes that were most likely to be found in rural Kenya. The evidence of this appears much later, but Burris, as has been said elsewhere, was strict about adherence to good technical practice, so we can assume that he was using these data-gathering methods from the outset. The design itself involves sizing calculations, and Burris developed simple processes for this, which would have been part of the training in the project. These sizing procedures certainly appear in Hankins' 1995 book. Installation involves a number of processes that would have been familiar to electricians but there are also procedures that are more specific to PV systems. For example, the commissioning of a battery – filling with electrolyte, its first charge, what to do if there is spillage of the electrolyte and so on. In operation, a PV system is straightforward but it performs better if a few simple energy-saving habits are cultivated, and the information supplied on the charge regulator is understood and its implications addressed. For example, if the regulator or indicator shows the battery charge to be low, then it is better not to use the loads until the charge returns to a high level again. The customer should be aware of these kinds of operational details, and so it would have been important to include this in the training. And, finally, maintenance of a system is simple but, again, important: cleaning the module, topping up the battery, checking connections are secure and so on. All these aspects are present in Hankins' book, and they were part of the training courses given elsewhere. So we can see that the project was also important as an early attempt to set an institutional trajectory. These procedures had to be articulated so that they could be expressed in the training, and Hankins acted as translator here between Burris and the technicians. Burris explained the technical details to Hankins, who then attempted to write these in a form that the technicians could understand.

The business impact of the three-schools project was similar to the Karamugi experience. Once a system was installed in a school, interest was stimulated among the local community and orders for systems began to flow. Here was more evidence that demonstrating the technology was a powerful marketing device. Further, as a later study showed, many people learned about PV systems and bought them as a result of seeing an example in a neighbour's house (van der Plas and Hankins 1998). As Hankins (2007, interview) puts it: 'Once someone had bought a system, he would have four or five friends come over and they would all want one too.' Again, we can identify learning but not necessarily whether it is of a first- or second-order quality. There is something of a first-order dimension to it in that learning that PV can supply electricity has an instrumental quality; that is, if someone wants to get access to electricity and then finds a way to do it, then that is instrumental learning. Whatever the quality of the learning processes, we can certainly infer that demonstrations helped to articulate socio-technical visions; those who were working with PV systems were able to conceptualise them, to a lesser or greater extent, in precise terms, communicate these terms and hence collectivise a socio-technical vision. So systems could be described: what they looked like, how much they cost, their functionality, their reliability, who could install them and so on. Information in this form is much more readily transmitted in conversation, enabling personal networks to act as effective communication channels.

In the same way that users and customers were able to describe systems in more precise terms, supply-side actors were able to articulate the market more precisely. By installing systems in homes, Burris, the technicians and others were meeting customers and developing knowledge of who they were and what they wanted. In other words, they were able to begin articulating the market with sort of learning-by-doing market surveys. We can assume the technicians would already have had considerable knowledge of local culture, including energy use, but it may not have been articulated in any detailed sense. Faced with having to explain PV systems to customers and how they would fit into their lives, in the hope of persuading them to buy, they would likely develop this articulation to some degree. Burris, also, is likely to have had some knowledge of local culture, having already lived in Kenya for many years and been married to a Kenyan. Still, we can assume that he would have learned a great deal in his interactions with customers and this wouldl have helped him to refine aspects of the technology as well as understand the market better. These more precise articulations would have informed marketing strategies as well as technical developments and socio-technical visions.

Development of marketing models in the household PV sector

The market began to grow quickly during 1985 and 1986, although figures for the number of systems installed are only estimates. Hankins (2007, interview) believes there could have been about a million dollars' worth of installations *altogether* over the ensuing two years (amounting to a few thousand systems at between USD 500 and USD 1000 each), with Solar Shamba doing many of these. Other estimates for Solar Shamba range from about 150 systems (Hankins 1990, p. 72), to 'hundreds of solar home systems' (Hankins 2003, p. 2), to more than 500 homes (Perlin 1999, p. 135), although this last figure is taken from Hankins (1987, p. 107) and seems to be a *total* for Kenya, as of January 1987, rather than entirely attributable to Solar Shamba. Duffy *et al.* (1988, pp. 3–5, Table 3.1) report that there were USD 218,000 worth of PV imports from the USA to Kenya in 1986. Up to and including 1986, the estimate is 82 kWp. The first year that we have an indication of module *sales* is 1987, estimated to be 88 kWp.

Prior to June 1986, there had been import duties and VAT on PV modules (Acker and Kammen 1996, p. 92). Import duties had been at 45 per cent but were completely removed because of lobbying by, it is claimed, the World Bank (Jacobson 2004, p. 142, n. 184) and the private sector (Acker and Kammen 1996, p. 92). Actually, according to Hankins and Bess (1994, p. 7), there was no official duty rate for PV equipment prior to the 1986 'removal'; any import duties that were applied depended on an arbitrary choice by the customs official at the border. There does seem to have been confusion, at the very least, over whether duties should be applied; Muchiri (2008, interview) states that modules would be categorised differently depending on whether they had diodes[10] attached or not.

Still, whether the imposition or removal of duties and VAT made any difference to sales is, according to Acker and Kammen (1996, p. 92), 'subject to debate'. They cite two, apparently opposing, views: that of Hankins and Bess (1994) and that of Karekezi (1994). Hankins and Bess (1994, p. 7) claim that the sales of modules 'increased dramatically', but Acker and Kammen (1996, p. 92) state that 'Karekezi found ... no savings were passed on to the customer.' Judging by the estimates reported in Figure 4.1, we can see that sales did rise very quickly in the period 1986 to 1988 but this could have happened for reasons other than price reductions. First, sales were starting from a low base and, second, this period was the beginning of intense marketing by a number of companies.

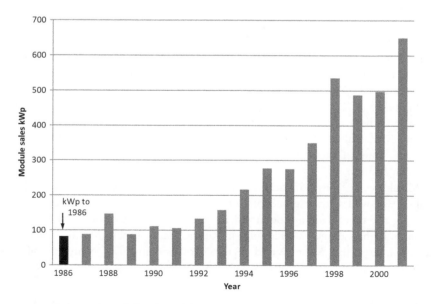

FIGURE 4.1 Estimated module sales, 1986 to 2001
Source: Hankins et al. (1997) and BCEOM et al. (2001).

The dealer network approach

At least one other approach was being developed at about the same time as Burris was building his business. Charles Rioba, a chemical engineer who had worked in the Biomass Department at the Ministry of Energy and Regional Development (MERD) from 1983, had become interested in solar systems and was looking for a way to develop his own career, the prospects for which he saw as unpromising in the ministry (Rioba 2008, interview). He took a year out from the ministry to do a master's degree in renewable energy at the University of Reading in the UK during 1984–1985, returned to the ministry and registered his own company, Solar World, but did not yet work on it full-time. Instead, he decided that he needed more practical experience and managed to get a job with Total Solar, a subsidiary of the French petroleum company Total, which had a network of outlets across Kenya.

Total were interested in selling PV in Kenya[11] and were looking to develop a business model. Rioba became their Dealer Development Manager in late 1985 (Rioba 2008, interview). Total Solar were mainly involved in solar thermal systems but, according to Hankins (1990, p. 67), they began to include PV in late 1985, around the time that Rioba joined them. Rioba spent his time setting up dealerships around Kenya, 'trying to identify risk-takers' (Rioba 2008, interview). Two of the marketing techniques he developed were installing subsidised demonstration systems in homes and setting up demonstration kits in the dealership outlets. These demonstrations, according to Rioba, were the most effective for persuading people to buy systems, especially the demonstrations in homes. Total Solar were not, initially, interested in installing systems in households – they were more interested in larger systems – but household installations became a more significant part of the business over time; according to Hankins (1990, p. 68) they installed about 50 household systems in 1986 and had installed about 550 systems by May 1990, by which time 'they [preferred] to install the kit themselves using the company's trained technicians' (Hankins 1990, p. 67). According to Rioba (2008, interview) and Masakhwe (1993, p. 67), Total trained their own technicians as part of the dealership package. These were short courses – about three or four days – and covered both solar thermal and electric systems (Rioba 2008, interview). Altogether, Rioba estimates that about 80 technicians were trained in this way, including Rioba himself, some of them working in the dealerships and others directly for Total Solar in Nairobi.

Masakhwe (1993, p. 67) acknowledges the importance of Total Solar and their training, as well as their pioneering of the dealer network approach to marketing PV. By the time Hankins conducted research on the sector for an MSc in 1990, Total Solar had about 12 dealerships in Kenya: Kitale, Embu, Mombasa, Kisii (the dealer here being Solar World, Rioba's own company), Nanyuki, Malindi, Eldoret, Kisumu, Nyeri, Meru, Nakuru and Nairobi (Hankins 1990, p. 67; Rioba 2008, interview). And, by 1990, other companies had embraced the dealer approach 'as most companies [could not] afford to competitively operate from Nairobi without local agents or dealers' (Hankins 1990, p. 69). Competition in the market had been increasing and, from about 1987, the companies had begun 'intensive marketing campaigns employing both the commercial media (newspapers, magazines and radio) and district agricultural fairs to advertise and demonstrate their products' (Hankins and Bess 1994, p. 3). There were various kinds of interactions between companies. Some of these were commercial: buying modules from each other, occasionally in quite large quantities (Hankins 1990, pp. 62, 66, 78; Rioba 2008, interview). Other interactions were more indirect, such as the movement of technicians between companies (Muchiri 2008, interview). Muchiri, himself, is an example. He trained and worked with Burris, moved to Alpa Nguvu, spent some time freelance and, at the time of being interviewed, was working with Rioba at Solar World. And it is well documented that many of Burris' other technicians went on to work with other companies or start their own businesses (Hankins 1990, p. 72; Hankins 1993, p. 33; Acker and Kammen 1996, p. 87; Perlin 1999, p. 135). Judging by the speed with which companies moved into the household market initially, and then used similar marketing and distribution methods, it is reasonable to assume that information and knowledge flowed quite freely between them.

This was a serious issue for Solar Shamba. Burris was known to make enemies of those he considered to be less technically conscientious than he was or, at least, those who did not practise to minimum technical standards (Hankins 2007, interview; Kithokoi 2008, interview). With the rapid growth of the PV market and increasing competition, many were finding ways to cut costs, and this was most easily done by omitting the charge regulator, using thinner wires, installing batteries of inadequate capacity or quality, including incandescent lamps instead of fluorescents and choosing modules of insufficient power output for the needs of the system. Burris tended to openly criticise those technicians, or others, who made use of any of these practices. As a result, Hankins (2007, interview) observes that:

> [Although] Harry [Burris] had put a business model in place ... he wasn't the type of person to attract business from an investor – that is, investors would not find him an attractive proposition. He was so adamantly independent. The business community in Nairobi steered clear of him and wouldn't invest in him and the technicians, except for the ones he worked closely with, didn't bring him business. They just started doing business on their own, the companies set up their own marketing channels, and left Harry out. Gradually, Harry was becoming isolated.

Burris left Kenya towards the end of 1987 or in early 1988 (Hankins 1990, p. 70; Hankins 2007, interview). Although Solar Shamba stopped doing business, Daniel Kithokoi, who had been working closely with Burris, started his own company – Solar Energy Installations – and continued to work in Meru, the area he had been covering while with Burris (Hankins 1990, p. 70; Hankins 2007, interview; Kithokoi 2008, interview).

PV as a consumer product

Two interesting developments occurred in the market during 1989. First, amorphous modules became available in Kenya (van der Plas and Hankins 1998, p. 298). Second, it seems that Chintu Engineering was given the licence to assemble kits using these amorphous (Chronar) modules and began supplying them separately, and as part of the complete solar lighting kits, from May 1989 (Hankins 1990, p. 63). Chintu supplied the modules and kits through its own three branches, a dealer network and, most notably, through Argos Furnishers, a very large company with over 30 branches in Kenya (Hankins 1990, pp. 64, 69). Argos offered the kits on a cash or hire-purchase basis – 'in the same way that they provide credit terms for bicycles, televisions and sewing machines' (Hankins 1993, p. 39) – the first time PV was available to the consumer on any credit terms (Hankins 1990, p. 64). It was already widely recognised, of course, that the initial cost of a PV system was high and that this could be a problem for the adoption of the technology, even if the lifetime cost could be competitive with other technologies. However, those supplying the household market in Kenya did not have the cash flow necessary to introduce hire-purchase, or other credit schemes, into their selling strategies. Hankins (1993, p. 39) notes:

> A shortage of credit for potential system buyers is the greatest impediment to expansion of PV sales. Many potential customers have steady incomes but are unable to amass the initial capital required to purchase systems. Local dealers cannot profitably offer credit because their own cash flow is limited and because of the problems associated with collection of debt.

It appears that Chintu was doing well on the basis of supplying these kits and selling them through Argos, as well as others. According to Hankins (1990, p. 64), the company sold 1200 modules in less than one year after introducing the kits (500 of them to Argos) and had assembled another 1000 kits by May 1990. However, the hire-purchase offering ended when Argos closed many of its rural outlets 'due to economic reasons' (Hankins and Bess 1994, p. 14). Those reasons are not given, but the period following the introduction of the kits was a difficult one in the Kenyan economy – a period that Acker and Kammen (1996, p. 90) describe as a 'two-year tailspin' – particularly after the suspension of quick – disbursing aid by donors starting in early 1992. Figure 4.2 shows the rapid increase in the consumer price index (CPI) and the fall in the value of the Kenyan Shilling against the US dollar, the CPI only really coming under control in 1995, even if the Shilling did not recover.

Socio-technical analysis of marketing model development

Total Solar and the dealer network

It is not entirely clear why Total moved into the PV market. According to Rioba (2008, interview) they were only doing this for 'greenwashing', perhaps a response

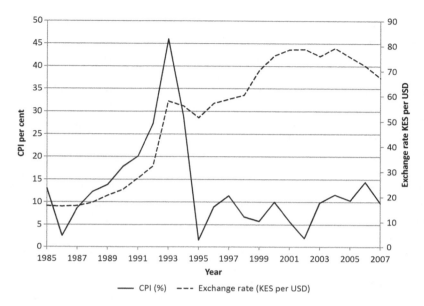

FIGURE 4.2 Consumer Price Index and exchange rate of KES to USD, 1985–2007
Source: African Development Indicators (2009).

to the growing environmental awareness worldwide. Nevertheless, they had most of the solar water heater market (70 per cent) and may have thought there was a sizeable donor market in PV worth pursuing. In time, however, it was the household market that became more important to their business. Whatever the explanation for Total Solar's involvement, the evidence does suggest that they were the first to develop the dealer network approach. And, over the next few years, other companies embraced this approach as the competitive pressures in the market intensified.

For Rioba, this was an important period of learning. He had gone to Total Solar purposefully to learn, and he certainly gained technical training, as well as the experience of setting up the dealer network. While doing this, he also gained useful experience of the market and how to sell to it, although this may have been more indirectly through dealers than directly through interactions with customers. He also saw the effectiveness of demonstration systems for generating business. Indeed, the dealers would have seen the importance of this strategy themselves.

It is difficult to identify whether Rioba's learning was of a first- or second-order quality. He may have experienced both kinds. We can be reasonably certain, however, that he had some form of expectation that guided his decision to join Total Solar. The source of this is likely to be a combination of the experiences he gained working in the MERD and studying renewable energies for his master's degree. Out of these experiences we could suppose that he formed a somewhat vague socio-technical expectation that incorporated renewable energies and business in Kenya. Given that his first degree was in chemical engineering, we can think of his forming of a personal socio-technical expectation as the result of second-order learning; he had changed his assumptions and was attempting to achieve a new goal. The learning that followed was concerned more with the detail of this expectation: technical details of PV, how to establish a dealer network, how to stimulate local markets and so on. His activities, then, began to articulate some of the detail and so helped him to form a socio-technical vision, in the Berkhout (2006) sense, but on a personal level. Some of this was collectivised by interactions with dealers and the installation of demonstration systems.

For other companies, the existence of a dealer network and demonstration systems were observable and, therefore, possible to imitate. Moreover, Total Solar appeared to be doing quite well in terms of business, and this would have served to demonstrate a market demand in more of the rural areas. We can see here a possible method by which Total Solar's business and distribution models could be copied, and a possible reason for companies wanting to copy them. However, apart from the fact that other companies adopted a dealer network approach, we do not have the evidence to conclude that they actually copied from Total Solar.

The dealer network that Total Solar developed was important for generating more business, of course, but it was also important for raising awareness of PV among more Kenyans. Likewise, the networks developed later by other companies had this effect. Further, the technical training that Total Solar conducted within its own network helped to establish at least some PV-specific capabilities around the country. While it is likely that this training was not as comprehensive as that given by Burris and Hankins (Rioba talks of three or four days to cover both solar thermal and electric systems), it was an attempt to institutionalise professional practice of a kind.

Chintu, Argos and hire-purchase

For market growth, the introduction of amorphous modules was important because they were significantly cheaper than the crystalline variety, even though the poor quality of the modules caused many problems. From the customer's perspective, however, a lower price was not the only benefit. The modules were rated at 12–14 Wp, which is a good match for a PV system that could power a portable TV. The modules began selling quickly, although it is difficult to know to what extent this was because of their size-price characteristics and to what extent it was because of the hire-purchase offering through Argos Furnishers. But the development was a significant articulation of market demand and, in terms of units sold rather than watts-peak, it soon became the most popular PV module on the Kenyan market (van der Plas and Hankins 1998).

We would expect that, once the hire-purchase option was demonstrated to be effective for generating business, the other companies would have introduced their own hire-purchase schemes. But this was a difficult process to manage. Argos already had plenty of experience with other products and so was able to include the kits relatively easily. For the other companies in the PV market, this would have been a risky venture that would have required setting up hire-purchase schemes, or some other form of credit facility, from nothing: no existing procedures and no prior experience.

So, while we can suppose that learning about consumer credit would have occurred among other players in the market, and perhaps created a desire to imitate such a facility, we can see that this was not enough to stimulate its widespread diffusion. Significantly more information, knowledge and experience were necessary (not all of which were observable) before other actors could adopt this approach. Moreover, the Kenyan economy went into a difficult period soon after this, causing Argos to close many of its outlets. It is reasonable to assume that, even if others were considering the introduction of hire-purchase at this point, the difficulties were too complex and the economy too weak to risk such a move.

Broadcasting the news: disseminating knowledge beyond Kenya's PV market

This section focuses on the dissemination of knowledge of the Kenya PV phenomenon beyond the actors directly involved in the market in Kenya. The first attempt at this is given in a book Hankins published (Hankins 1987). It was his first book and covers renewable energy in Kenya in general. While there is a chapter on solar energy, there is only about half a page on the PV market specifically. In this, he could point to just a few hundred systems installed, and so it would be difficult to persuade anyone that there actually was a phenomenon. Still, there are some other aspects of this first attempt to disseminate that might be important in terms of the development of a PV niche in Kenya. First, Hankins had to do the research. That meant travelling around Kenya to various projects, and so he would have been able to network far more extensively than he had done before this. The book was paid for by USAID and, latterly, the Canadians. So, Hankins was building a reputation among some of the donors that would be helpful to him later.

Dissemination and recruitment

Hankins left Kenya towards the end of 1987 and returned to the USA. He struggled to find the means to get back to Kenya but was certainly trying. Eventually, he went to do his MSc at Reading (the same master's degree as Rioba) in 1989. For this, he went to Kenya to do his fieldwork in 1990. He then discovered that the market had flourished since he had left. He did a survey of a number of PV systems and wrote this up for his dissertation. The 'message' in this was 'picked up by the World Bank' (Hankins 2007, interview). This time, although it may still have been a modest phenomenon, Hankins had very detailed descriptions of the uses of PV systems in rural areas of Kenya, some of which were for 'productive uses'. He had also captured some of the local practices, good and bad.

Hankins did other research, including a trip around Eastern and Southern Africa, during which he met many players involved in PV. He also did research for another book, this time funded by the Solar Electric Light Fund (SELF). In late 1991 or early 1992, he teamed up with the NGO Kenya Environmental Non-Governmental Organizations (KENGO) in order to organise a regional workshop on PV, an idea that Hankins and Burris had conceived. He had put a proposal to the African Development Foundation (ADF) in the USA (having been encouraged by a contact there who was ex-Peace Corps). Hankins invited many of the contacts he had made during his trip around Eastern and Southern Africa to attend the workshop.

The workshop was held in Nairobi in March 1992 and was attended by people from across East and Southern Africa, including some from the MOE in Kenya. (Actually, it was a broad selection: private sector, NGOs, government, universities, donors, individuals.) The format of the workshop included formal presentations, training content, and practical work to install a PV system in a rural area (Meru, where Burris had been working). For Hankins, it was highly successful. He had two or three immediate possibilities for projects that came out of the workshop. In order to get the funding for these, he had to work through a legally registered organisation and so started Energy Alternatives Africa (EAA) with Daniel Kithokoi (who had been working with Burris prior to this and had started his own solar business following Burris' departure from Kenya). Hankins also claims that this was the time when SolarNet[12] was started, although it was an unofficial organisation at this point and had no funding. One of the projects was to set up a solar training centre at Karagwe Development Association (KARADEA), located in the north-west of Tanzania (an area that is very difficult to access). KARADEA was being run by Oswald Kasaizi, who had attended the Nairobi training and discussed the project idea with Hankins.

The KARADEA project proposal was developed during a visit by Burris and his wife late in 1992. Hankins did not attend at this time. However, this was also when the Global Environment Facility (GEF) was about to start a PV project in Zimbabwe, and Burris was appointed chief technical advisor to it. He was, therefore, unable to pursue the KARADEA project, and Peter de Groot of the Commonwealth Science Council (CSC) brought Hankins in instead. Kasaizi and Hankins then put the proposal together for what became the KARADEA Solar Training Facility (KSTF), which the CSC funded. It involved a building that included a classroom, PV equipment and other facilities for training PV technicians.

Hankins also wrote a textbook on PV design and installation and got this published in 1991. He then updated it, and this was published in 1995. These texts were used in the training courses run at KSTF and elsewhere, thereby becoming important for institutionalising best-practice for PV design and installation. He also wrote a book in 1993, funded by SELF, covering four country case studies (see Hankins 1993), and teamed up with Mike Bess to write an Energy Sector Management Assistance Programme[13] (ESMAP) paper (cited above: Hankins and Bess 1994). These two, in particular, helped to articulate the Kenya PV phenomenon more widely than Kenya.

Socio-technical analysis of dissemination and recruitment

This was a highly active period for building networks and disseminating experiences. Hankins' primary immediate objective was 'to get published and to write a book' (Hankins 2007, interview). He used his experience of working on PV installations as a basis to further this objective; he formed something of a personal expectation or vision, where the goal was to get published and the means included writing about PV. His first opportunity to realise this came with the 1987 book on renewable energy in Kenya, for which he had to conduct extensive research around the country. He included a short section in the book on the current state of the PV sector, but this was a straightforward list of the numbers and types of systems installed in Kenya.

His research enabled him to network much more than he would have done prior to the book. The book covered most renewable energies and so could not treat any one of them too deeply. Even so, Hankins was able to learn a great deal about the extent of the PV sector in Kenya and to establish contacts in addition to those he already had through his work with Burris. Hankins had already formed a personal socio-technical expectation about PV in Kenya, and now he was able to start refining this into a vision through the learning he was doing in his research. This learning would most likely have been of a first-order quality: the kinds of systems in operation and their locations, who was working with the technology, the extent and nature of successes and failures and so on. Some of these details were included in the book but it was, for the most part, a catalogue of the state of renewable energy in Kenya. As such, it was a useful means for wider dissemination.

More significant, however, was Hankins' MSc dissertation. This was focussed exclusively on PV in Kenya, and articulated in considerable detail both the supply and demand sides of the market. He documented how the supply chains were working and how people in rural areas were actually using the technology, sometimes for productive purposes but mainly to improve the immediate quality of their lives. He learned about some of the problems in the market, some of which were technical issues and some to do with user practices. Here was an opportunity for him to persuade donors that there was a phenomenon worth encouraging – one that aligned with their institutional interests – but one that needed support, and, therefore, it was an opportunity for Hankins to find work in Kenya.

In his 1987 book, Hankins had already started expressing a socio-technical expectation of PV in Kenya. Using his MSc dissertation (Hankins 1990), he was able to strengthen his 'bid' by referring to 'thousands of systems installed through a private market' rather than the 'hundreds' that were in place during the period of the research for his first book. This apparently private sector phenomenon allowed him to connect with the increasingly dominant free-market paradigm that framed much of development thinking. He could point to the 'success' of the Kenyan PV market in diffusing an environmentally benign technology, which also supported development goals, while highlighting ways that it could be improved, in terms of scale and quality, through donor intervention – essentially, a market-failure argument.

Whether these arguments – this socio-technical vision – formed the basis of his early proposals is not possible to say, but he certainly framed his later descriptions in this way.

Hankins was certainly successful at attracting funding, and much of this enabled him to develop networks, both inside and outside Kenya, through which he could disseminate and develop a PV socio-technical vision. The money he received from the Canadians helped him to make contacts across East and Southern Africa, many of whom participated in the Regional Workshop in Nairobi in 1992. That event was, of course, both a networking and learning opportunity for the participants, but it was also during this time that Hankins was able to mobilise resources for projects in the region. In order to make use of these opportunities, he started EAA together with Daniel Kithokoi, and their work helped the company to become the most important PV actor in the region.

One of those early projects was with KARADEA to help establish KSTF, the first specialised PV training centre in East Africa. The relationship between EAA and KSTF persisted for about ten years during which 175 PV technicians, mostly from East Africa, were trained at the facility (KSTF 2009). Over that period, at least five donors supported the work: CSC, Sida, APSO,[14] Hivos, and the Ashden Trust. So network-building was extensive through the KSTF project. And the project maintained a space in which the basic PV training course could be developed and refined. Indeed, the KSTF course was something of a model for other courses conducted later in the region and informed later national PV curriculum development. So the KSTF collaboration was important in institutionalising PV practices in East Africa, developing networks, and collectivising a PV socio-technical vision.

We can see that this period was important for network-building and dissemination. While these have continued, it appears that it was here that the dominant form of the Kenyan socio-technical vision was refined and collectivised, communicated in the 'private sector-led development' narrative. Hankins was influential throughout this process and has expressed this vision in his papers, books, proposals and reports, as well as in training courses and other networking events. It has been a persuasive vision because of the fact of the rapid growth of the Kenyan PV market. Hankins, and others who followed, have accentuated the private sector aspect of PV market growth in Kenya and downplayed any donor influence. The somewhat ironic effect of this has been to convince a wide range of donors to fund interventions and other activities. These resources facilitated the early network-building, dissemination and training that were important for the learning, collectivising of visions and embedding of practices that helped to stimulate and sustain the growth of the PV market.

Articulating the market

This section investigates a number of product design and development, as well as market research, activities that occurred in Kenya, beginning in 1994. These activities can be categorised into two broad themes: (1) product design and development;

and (2) market surveys. A third category – PV cell research and manufacture – was the focus of activity in the Physics Department of the University of Nairobi, but was largely disconnected from the local commercial sector while being networked with international academic research. There is little to say other than the core interest of this work is in 'wet cell' research, something that is not yet commercial and does not seem to have had any appreciable impact on the Kenyan PV niche. The former two categories of activity were conducted by commercial actors – but funded by donors – and have helped to articulate the Kenyan PV market in their own ways, in sometimes fine detail. Due to space constraints, the research and development activities investigated here do not constitute all those that have occurred in Kenya. Instead, they constitute a sample (see Table 4.1) that serves to highlight the co-evolutionary dynamics on which SNM theory focusses, as well as being of relevance to the development of the socio-technical niche.

Controlled experiments

As we have seen, soon after EAA was formed, they began to implement projects in the region. One of the earliest projects was conducted in Tanzania (the KARADEA Solar Training Facility) but, in 1995, they began a solar lantern project in Kenya and there followed a long period during which they managed many other PV-related projects in the country. This section describes four technology projects they implemented, but also includes some description of the activities of Leo Blyth, who came to Kenya from the UK searching for a way to disseminate what were called at the time micro-solar kits, a variant of PV products that has come to be called pico-solar.

It is clear that, as early as 1990, Hankins was interested in the market possibilities of solar lanterns, although he considered them to be too expensive, too constrained in functionality, and difficult to repair locally (Hankins 1990, p. 80). Even so, he saw their potential to bring electrical services to a poorer segment of the population, and to do so as engineered systems rather than the *ad hoc* 'systems' that were becoming common in the market[15] (Hankins 1996, pp. 8–9). Through EAA, and

TABLE 4.1 Selection of projects and market surveys

Product design and development	Years	Market surveys	Years
Solar lantern test marketing	1995 to 1996	Survey of 410 SHSs	1996 to 1997
Micro-solar[a]	1996 to mid-2000s	STEP (Solar Technician Evaluation Project)	2000 to 2001
Jua Tosha battery	1997 to 1998	Survey of East African PV markets	2002 to 2003
Battery pack	1997 to 1999		
BOS components	1999 to 2001		

Note: [a] Micro-solar was the term used for what later became known as pico-solar.
Source: Based on Byrne (2011, p.107).

funded by SELF, he had already worked in northern Tanzania to supply a few batches of lanterns (OSEP 1998; Byrne 1999, p. 13). The Kenyan project, however, differed from the Tanzanian experiment in that the lanterns were placed in a sample of rural shops rather than being supplied through an NGO. This was a more market-friendly approach than the first lantern project and marked the beginning of a method that EAA used in many subsequent projects.

From those already available on the market, six models of lantern were selected for test-marketing and a seventh, prototyped by EAA themselves, was added (Hankins 1996, p. 11). These were supplied to six dealers: five in rural areas around Mount Kenya and one in Nairobi (Hankins 1996, p. 14). EAA tested a sample of the lanterns in-house and later questioned 65 per cent of those who bought lanterns, as well as asking the dealers for their opinions. There was a range of findings related to technical issues, functionality, consumer practices and preferences, the impact of taxes on price, supplier needs and some suggestions for ways to strengthen the marketing of lanterns.

The technical issues concerned the quality of the designs and how these might be improved. Functionality included recommendations for powering a radio as well as a light. Consumers were found to be conservative in their purchasing, especially the lower-income groups; the best-selling lantern had a shape similar to a pressure lamp; middle-income groups tended to buy the lanterns first, while lower-income groups were less likely to take risks and consumers did not like the monochrome light of LEDs. Taxes were seen to add about 30 per cent to the price of lanterns because they were categorised as lamps rather than PV systems (PV modules were taxed at either 10 per cent or 0 per cent). The needs of suppliers included access to a small range of standardised spares, which was also seen as a way to overcome some of the risk-averse behaviour of customers who would not buy a lantern unless spares were available. And three marketing methods were suggested. One, lanterns could be supplied in two stages: the customer would buy the lantern first, then pay for it to be charged until the cost of the module had been collected, upon which the customer would then receive the module. Two, a new product could be introduced which consisted of a battery and charge regulator combined into a single unit. The battery could then be recharged using a battery charging service, and the customer could get access to electricity while saving to expand their 'system' to include a PV module and better lamps later. Three, hire-purchase or other financing schemes could be used to help customers buy solar lanterns (Hankins 1996, pp. 31–36).

EAA managed to secure funding for projects to pursue two product ideas they had suggested in the solar lantern report. One of these was for a small locally-manufactured 'solar' battery, or 'Jua Tosha' as it became called; the other was for a 'BatPack', the battery and charge regulator unit mentioned above (Hankins 1996, p. 36). Both projects got underway in 1997. For the BatPack, the Ashden Trust funded EAA and ApproTEC to develop a prototype and this was ready in 1998 (EAA 2001, p. 5). The Jua Tosha project, supported by ESMAP, began in June 1997 and the first of a total of 800 batteries manufactured by AIBM were being shipped to up-country retailers by November (Ochieng *et al.* 1999, p. 9). Production of the BatPack – this second phase of the project being funded by ESMAP – did not get underway until April 1999, and test-marketing started in November (EAA 2001, p. 13).

The Jua Tosha project was largely successful: the battery was well received by the market, and dealers, who had not considered a 20 Ah battery[16] necessary, now wanted to see continued production (Ochieng *et al.* 1999, p. 27), and by the time of the report (August 1999), more than 200 units per month were being sold (Ochieng *et al.* 1999, p. 2). Also, one of the other battery manufacturers – ABM, who had first introduced a 50 Ah solar battery in 1985 – started production of a 40 Ah solar battery[17] (Ochieng *et al.* 1999, p. 27; EAA 2001, p. 4).

However, the BatPack project was considered unsuccessful in terms of its original objectives (EAA 2001): the product was unattractive to its target market, few units were sold and there were unresolved technical problems with the charge control unit. Eventually, EAA decided to import a similar product – the Sundaya Battery Pack (from Indonesia) – and test-marketed this instead, beginning in January 2000 (EAA 2001, pp. 8, 14). Apart from the BatPack's technical problems with the Rodson[18] controller, it was discovered that an investment of around USD 15,000 for a mould would be required if the casing were to be made from plastic – a large risk for a small Kenyan company, considering that thousands of units would have to be sold to recoup the investment (EAA 2001, pp. 8, 6).

While the BatPack report states that the project was unsuccessful, other aspects were highlighted in an attempt to suggest that the project had achieved some positive outcomes. One of these outcomes was an identified demand for this type of product, albeit among a higher-income group than anticipated and for a higher-specification unit than the one tested, unless the price could be reduced sufficiently. Evidence to support this claim included the observation that other suppliers, who were not involved in the project, began sourcing similar but higher-specification products from outside Kenya (EAA 2001). Another success claimed in the report was that Rodson were said to have introduced two new products to the market as a result of their involvement in the project: a charge controller and a battery monitor (EAA 2001). While it was probably fair to say the BatPack project inspired these product ideas, the work to design and develop the products appears to have been done through a project funded by MESP, beginning about September 1999 (Osawa 2000). Reflecting on these various projects, Osawa (2008, interview) remembers that there was a period during the early 2000s – up to about 2006 – when local manufacture of BOS components was very successful; indeed, EAA (2001, p. 2) reports that Rodson were selling 'several hundred' battery monitors and charge controllers per month. However, local manufacturing of BOS components has almost disappeared as a result of Chinese-made products coming onto the Kenyan market (Osawa 2008, interview).

For all these projects, EAA used a similar methodology. They persuaded up-country dealers to stock the prototype in their shops, waited for a period and then questioned the dealers and customers about their experiences with the product.

They also tested the product themselves, either in-house or with the help of an independent actor, to document the technical specification.

However, in the BOS components project, they introduced a new aspect by including focus groups with consumers and, separately, with dealers *before* the prototypes[19] were manufactured. The results of these focus groups informed the choices of products to manufacture and refinements to the designs of those chosen. The BOS project had initially proposed six product concepts, and the two that appeared to meet the most immediate market demand – the battery monitor and the charge controller – were the ones developed by Rodson (Osawa 2000).

A number of technical and functional issues were raised during the consultation and test phases of the product development (Osawa 2000). First, Rodson were requested to reduce the value that was considered a full battery charge, so that the full indicator would be illuminated for longer, providing a 'better' customer experience. Second, Rodson were requested to lower the value set for the low voltage disconnect so as to provide electrical services for longer. Third, Rodson were asked to introduce a reset button that would allow a few minutes of electricity supply once the low voltage disconnect had activated, giving the user light while they set up a kerosene lantern, for example. And, fourth, the charge controller was modified after it was discovered that it could not cope properly with inductive loads such as fluorescent lamps.

Another important feature of the way in which EAA worked throughout these projects was the extent of their networking. Between the four projects discussed here, they interacted with at least 39 different dealers and suppliers in 16 cities, towns and villages around Kenya, and at least five of the dealers were involved in more than one project (Hankins 1996, p. 14; Ochieng *et al.* 1999; Osawa 2000; EAA 2001, p. 31). These numbers do not include the manufacturers, donors and other organisations with whom EAA worked: AIBM, Chloride Exide and Rodson; ESMAP, the Ashden Trust and MESP; and ApproTEC, ITDG, SolarNet and the University of Nairobi Physics Department.

Finally, it is interesting to say something about the activities of Leo Blyth. After his first visit to Kenya in 1996, Blyth spent a number of years moving back and forth between Kenya and the UK, trying to disseminate DIY Solar[20] kits when in Kenya and finish a development studies degree in the UK. His dissemination efforts included training people to make the solar kits, conducting dozens of such courses in Kenya and other countries in the region and with various groups including Trans World Radio, SolarNet and the Peace Corps (Blyth 2008, interview). One such course was conducted in the Nairobi slum Kibera in June 2004 (Keane 2005, p. 7), and it may have been here that Fred Migai, who has so far been the only Kenyan to try to commercialise the idea (Blyth 2008, interview), learned to assemble the kits.

However, Blyth himself tried to commercialise a product idea around 2002 with funding from the Shell Foundation, developing the idea out of his experiences in the region with these 'pico-solar' kits and other products he had seen. He had also shown a few of the Chinese pico-solar products that appeared on the local market to Hankins, who liked the ideas but was concerned about the quality (Blyth 2008, interview). For the Shell Foundation project, he worked with EAA and used the BOS project methodology as a template. Following focus groups with consumers, a product to charge a mobile phone and power a radio was chosen and the project was to get 1000 units manufactured in China. However, the manufacturer 'ate the money' and the project collapsed (Blyth 2008, interview). Migai went on to assemble a simple kit that could charge a mobile phone and power a radio, although it had no charge controller or battery, and to sell the kits up-country himself and through a network of agents (Migai 2008, interview). Before he learned how to assemble the pico-solar kits, Migai had been a marketing agent for Swiss Guard, selling a healthcare product in Kenya through a pyramid marketing scheme (Blyth 2008, interview). It appears that the methods he used to sell the pico-solar kits were similar to those he practised while working for Swiss Guard and, at the time of the interview, he claimed to be selling around 100 solar kits per month (Migai 2008, interview).

While Blyth continued for some time to try to commercialise pico-solar products in Kenya, he later considered local manufacture to be the wrong direction – it takes large amounts of investment and needs large volumes to be viable; otherwise, the transaction costs are too high (Blyth 2008, interview). Indeed, Osawa (2008, interview) came to the same conclusion regarding manufacture in Kenya. It is interesting to note that no actors, other than Blyth and Migai, were interested in pico-solar products at the time. Indeed, apart from Hankins' general interest, many were deeply sceptical. But, as we will see in Chapter 5, pico-solar lanterns have become a highly successful product in Kenya.

Market surveys

There have been several surveys of the Kenyan PV market and, as we would expect, they have served to articulate and codify many aspects of it. We have already considered two of these in relation to the dissemination of the PV phenomenon in Kenya. Both of these were conducted by Hankins (1987, 1990): the 1987 survey was more of a cataloguing project, while the 1990 survey investigated some of the detail of the demand and supply sides of the market. Since then, there have been at least eight surveys that have focussed on SHSs in the PV niche in Kenya. Numerous other studies have been conducted, but they have either incorporated PV into a larger survey or they have not been surveys. Of the eight that were focussed on the PV niche, one is unavailable[21] (Musinga *et al.* 1997). Consequently, the discussion here is based on the other seven surveys: Hankins and Bess (1994), Acker and Kammen (1996), Hankins *et al.* (1997), Jacobson (2002a; 2002b; 2004) and ESD (2003).

The Acker and Kammen survey was conducted in July and August of 1994 and included, among other aspects, interviews with 40 owners of PV systems sized between 10 Wp and 100 Wp (Acker and Kammen 1996, p. 93). It asked similar questions to Hankins' 1990 research and found similar benefits and problems. In this sense, it supported Hankins' work and further elaborated his initial articulation of the market: who was buying systems, the kinds of systems, how they were being used, how consumers learned of PV, consumer expenditures, performance of systems, typical benefits and problems and (an addition to the information gained by Hankins) the distance to the grid.

Some of the more surprising findings of the survey included the discovery that PV systems were being bought by people who could not be considered affluent, and some appeared to have struggled to acquire their systems (Acker and Kammen 1996, p. 95): 'Many of the households whose annual incomes are less than the survey average of USD 2800 are spending over 75% of their income for their systems, with some homes spending almost 200%.' Indeed, a visual inspection of one of the graphs in the document suggests that up to a quarter of the systems investigated in the survey were bought by people who had an annual income of less than USD 1000, and a few of these systems cost more than USD 1000 (Acker and Kammen 1996, p. 96, Figure 23). The understanding up to this point was that reasonably well-paid consumers, or cash crop farmers and other business people, were buying systems (Hankins 1990, p. 3; Hankins and Bess 1994). Another interesting finding was that a quarter of the systems were in homes within 1 km of the grid – in partial support of an estimate of 40 per cent given in Hankins and Bess (1994, p. 5) – even though the 'break-even distance beyond which PV would be cheaper' was estimated to be 8.8 km, and that one of the systems was in a home actually connected to the grid (Acker and Kammen 1996, p. 96).

One of the questions not asked was whether, and how much, savings were sustained as a result of using PV systems. The next survey of household systems investigated this question, along with many of the same dimensions addressed by the Acker and Kammen study. The survey, funded by ESMAP, was conducted through EAA from December 1996 to March 1997 and covered 410 household systems in 12 districts across Kenya (Hankins *et al.* 1997, p. 2), forming the basis of an *Energy Policy* paper written by Robert van der Plas of the World Bank and Mark Hankins (van der Plas and Hankins 1998). The savings which the survey found were most significant for smaller systems and, overall, the majority of savings were on kerosene and dry cells, equally shared (Hankins *et al.* 1997, pp. 37–38). The significance of the savings enjoyed by those with smaller systems was heightened because there appeared to be a trend in the market toward smaller systems, already indicated, to some extent, in the Acker and Kammen study (Acker and Kammen 1996, p. 97, Figure 26), facilitated by the availability of 12 Wp amorphous modules. The average savings were about USD 10 per month and, for those with systems smaller than 15 Wp, USD 8.55 (mostly on dry cells but also on kerosene and battery charging) (Hankins *et al.* 1997, pp. 36–38).

Other than these findings, the survey was generally in line with the findings of the previous studies but, of course, the number of systems investigated made it an important articulation of the market. And this enabled Hankins *et al.* to present detailed recommendations assigned to all types of actors with an interest in the market: government, donors, industry, financial institutions, NGOs, and research organisations (Hankins *et al.* 1997, pp. 47–53). There were also recommendations made from the Acker and Kammen survey, and the Hankins *et al.* study over-lapped with these in a number of ways: the need for supportive policy, both national and international; the need for capacity building; that finance schemes should be introduced; standards and codes of practice should be developed to overcome the quality problems; there was a need for better and impartial infor-mation and there should be smaller engineered systems, such as solar lanterns, and more modular provision of system components in the market (Acker and Kammen 1996, pp. 105–108; Hankins *et al.* 1997, pp. 47–53). It is interesting to note that the Hankins *et al.*'s recommendations made a point of insisting that subsidies were not to be used to promote PV systems (Hankins *et al.* 1997, p. 48): '"Project" (public sector) funds should be channeled in ways which will grow the market, without subsidizing systems or Government institutions.' This was in line with the report's general assessment of the PV market in Kenya's being a private sector phenomenon. In the Introduction to the report, it states that there had been an 'absence of Government, finance or donor support – or any project intervention effort' in the Kenyan PV market, acknowledging only that '[s]everal independent volunteer initiatives were instrumental in catalyzing the existing market, but these were neither expensive nor large scale' (Hankins *et al.* 1997, p. 9, n. 2). This is interesting, given the record of donor-support we have reported above, but also because there were later changes to this position, at least on the part of ESD[22] who began talking of 'smart subsidies' (ESD 2003).

Jacobson, with the help of others, and working through EAA (which later became ESD), conducted a number of surveys between 2000 and 2004 (sample size in brackets): Solar Technicians (366); Solar Vendors (312); Solar Households (76); and Energy Allocation (15 households) (Jacobson 2004, pp. 302–309). For the Solar Technician (STEP – Solar Technician Evaluation Project) and Solar Vendor surveys, Jacobson employed two local technicians to conduct the majority of the fieldwork: Maina Mumbi and Henry Watitwa (Jacobson 2002a, p. 7). For the household study, Jacobson employed the same two technicians to conduct many of the interviews (Jacobson 2004, p. 304). The energy allocation survey involved using data logging equipment to measure appliance use over a period of four to six months for each system and was supplemented with ethnographic observations (Jacobson 2004, pp. 306–307).

The technician and vendor surveys were important because they characterised the supply side of the market more thoroughly than had been achieved up to that time. The main findings from the technician survey were that most technicians (90 per cent) operating in the PV market were not solar specialists, and only 5 per cent of solar technicians had regular employment in PV services (Jacobson 2002a, pp. 9, 11). Similarly, the vendor survey discovered that only 5 per cent of shops stocking PV equipment were specialists, and 41 per cent were hire-purchase shops (Jacobson 2002b, p. 31).

The most important conclusion that Jacobson drew from these findings was that PV training courses needed to be re-designed to be shorter, delivered in up-country locations and targeted to the needs of non-specialists who were, nonetheless, working in the PV market (Jacobson 2002a, p. 8). This was a departure from the form in which EAA had been conducting their training courses for many years, developed from the three-schools project in 1985 and the work at KSTF since 1993. Whether it was a result of the study or not, the training courses supported through the Photovoltaic Market Transformation Initiative[23] (PVMTI) from 2006 seem to have been arranged according to Jacobson's recommendations to some extent, particularly the delivery of courses up-country and the targeting of non-specialists (Nyaga 2007, interview; PVMTI 2009). And an interesting impact of having employed two local technicians to conduct the majority of the interviews was that their interactions with so many other technicians stimulated discussions of forming their own association: the Kenya Solar Technician Association (KESTA) (Watitwa 2008, interview). Although KESTA was officially registered in 2005 (SolarNet 2005, p. 28), it did not attract funding or manage to collect subscription fees and so was unable to achieve much for technicians (Watitwa 2008, interview).

The two other surveys conducted by Jacobson during 2003 and 2004 provided insights into the dynamics of electricity use within the household. Although the sample was very small in the energy allocation survey – just 15 systems – the detailed information of appliance use, combined with observational material and interviews, provided evidence of a more complex reality of electricity consumption patterns in the home than was previously available. It was assumed that electric light benefited women and children, reducing their exposure to kerosene fumes in the kitchen and improving conditions for studying at home. Or, at least, this was the rhetoric within the development regime in regard to connections between electricity and development. Jacobson's survey discovered that this was not necessarily the case, particularly in households with small systems. He found that TV dominated electricity consumption in homes that had a small system (less than 25 Wp), using 54 per cent of the energy available, and that the kitchen often had a low priority when deciding where to install lights; for larger systems the majority of energy consumption was for lights (61 per cent), with TV accounting for one-third of consumption (Jacobson 2004, pp. 204–232; 2007, pp. 153–155). It is unclear whether these findings have had any impact on the rhetoric on PV and development; it may be too soon to be able to note any effect.

The most geographically wide-ranging survey of the PV market to this point was conducted for the World Bank through ESD in 2003, covering seven countries in Eastern Africa. One of the stated aims of the study was to be able to describe the development of PV markets in the region. The results showed quite different kinds of markets across the countries studied, with Kenya clearly the largest and most developed, described as 'mature' (ESD 2003). It updated some of the fundamental information about the market such as installed capacity, but also provided statistics on numbers of companies and technicians operating, described increasing complexity in the supply chains and marketing strategies, and gave figures for awareness of PV among the population. Above all, however, it gave a detailed and highly prescriptive set of recommendations on how to develop PV markets in the region including, for the first time, some support for the use of subsidies in PV promotion, argued on the basis that PV markets had been stimulated to grow rapidly in some of the industrialised countries through the use of subsidies (ESD 2003).

Socio-technical analysis of market articulation

Controlled experiments

It is clear that the implementation of these various projects generated deep inter-actions between actors from different sectors and throughout the PV supply chain within Kenya. Further, the projects provided opportunities to learn a great deal about both the supply and demand sides of the PV market: about user practices and preferences, supply-side practices and assumptions, technical details of product concepts and formal institutional constraints such as VAT and other taxes. We can also see that there was important system-building work being done by some actors, EAA being perhaps the most significant of these. Hankins, in particular, appears to have developed a proposal model that succeeded in aligning the interests of the development regime and the needs of actors within the Kenyan PV niche, while linking to others, such as battery manufacturers and electronics specialists. By deploying a socio-technical vision in which PV diffusion could be achieved through the private sector, he was able to attract resources for experimentation that the private sector would have found too risky to provide, but from which it benefited significantly.

The learning generated by these experiments resulted in better articulation of the rural market in two senses: a clearer description of its characteristics, and a strengthening of interconnections between actors in the supply chain. In turn, this better articulation, in both its senses, helped to enhance and collectivise expectations of market demand. Equipped with a richer understanding, actors changed their behaviours and introduced new products to the market, guided by finer-detailed socio-technical visions. This is not to say that the projects were straightforward, consensual and positive in all their aspects; there were technical problems, negative outcomes and, at least where pico-solar products were concerned, the *suggestion* of dissensus.

Where technical problems were discovered, their solution was generally the result of first-order learning; for example, the modification to the Rodson charge controller so that it could cope with inductive loads (BOS components project) and the sourcing of a product similar to the battery pack when the Rodson control circuit could not be made to work (BatPack project). It was also through first-order learning that expectations were developed into visions: more precise detail in various aspects such as consumer demand, consumer practices and preferences, willingness to pay, product functionality and quality, local manufacturing capacity and the impact on price of taxes. This filling-in of details was important for niche actors because it lowered the risk of investments for them; they had better information about the market and their role in it, enabling them to articulate business models.

In regard to negative outcomes, it is interesting to observe that these were a source of second-order learning. For example, the lack of demand for the battery packs challenged assumptions that shifted actors' expectations. The shift, in this case, was from targeting a poorer segment of the population to a wealthier one. At the same time, the challenge to assumptions generated a new understanding of the preferences of the poorer segment: that functionality and price are far more important than convenience. And Blyth appeared to adjust his expectations about the means to achieve greater diffusion of pico-solar products, based on what we might characterise as the negative outcomes of his experiences working with NGOs – the disappointing adoption rates for the solar kits he demonstrated.

However, unlike the private actors in the other projects we have discussed, it took a long time for Blyth to realise this shift in expectations. The explanation for his persistence could lie partly in his personal expectation that PV was 'not just another product' (Blyth 2008, interview) and therefore could 'win' on its own terms, and partly in the examples of two actors who did not see PV in this way. Instead, at least one of them – Migai – marketed pico-solar in a similar way to other small products and achieved some success. Here, Blyth was presented with an alternative vision, at least in terms of the means by which pico-solar technology diffusion could be realised, and it is one he appears to have assimilated. This suggests that second-order learning can occur as a result of positive outcomes as well as negative, but, in this case, it occurred through observation of the positive outcome for others – a kind of vicarious second-order learning. Indeed, we can see that something similar occurred with the Jua Tosha battery (the other battery manufacturer in Kenya, who was not involved with the project, introduced its own small solar battery soon after the project finished) and, in some ways, with the BatPack project (another supplier, again not involved with the project, sourced a similar product, even if this did not result in any market penetration).

Second-order learning *opportunities* may also have existed as a result of the dissensus over pico-solar products. It is not entirely clear from the evidence, but we could reasonably argue that the paucity of experiments with pico-solar products, and the consequent lack of assumption-testing, constituted one source of this dissensus. There was some testing going on, but it was not being documented or studied systematically; Chinese companies were trying various products in the market, and Migai was selling units through a network of individuals. The only project that would have provided some documented testing of assumptions was that funded by the Shell Foundation, but, as with many projects that are considered failures, documentation was difficult to find and few actors wanted to discuss it. Nevertheless, the characteristics of pico-solar products appear to be aligned closely with practices and preferences among consumers in Kenya, and so we might expect the products to be easily embedded in the market. Further, the movement in the market has been toward smaller systems, and much of the motivation for the projects described above has been to enhance the technical quality of such systems. Moreover, there is growing interest in the 'Bottom of the Pyramid' approach to development, and pico-solar appears well aligned with this expectation.

Given these conditions, it is difficult to understand why the pico-solar 'market' did not attract much interest from the established PV actors in the region or from donors. On the contrary, many of the established PV actors held negative expectations about pico-solar; only a few actors held positive expectations and used these to guide their activities. These pico-solar 'promoters' were working almost entirely in the private sector with meagre resources and independently of each other. However, having said this, the International Finance Corporation's (IFC) *Lighting Africa* project, which began operation in 2007, may be an indication that the situation was beginning to change, at least as far as lighting products were concerned (World Bank 2007). Blyth became a consultant to the *Lighting Africa* project (pers. comm., Blyth) and, as we will see in Chapter 5, the pico-solar market expanded rapidly from 2009.

Finally, it is important to recognise that these projects involved many of the same actors and that there was a consistency or stability in the networks of actors. Certainly new actors joined and not all the actors participated in all the projects. Nevertheless, this relative stability facilitated the building of trust (this has been important for eliciting information for market surveys) and the accumulation of knowledge generated in the projects. Moreover, EAA has been a central actor in these activities, as well as many other projects not considered here. This has been important for at least two reasons: first, it has enabled EAA to be a cosmopolitan actor in the local PV niche, in the sense used by Deuten (2003), or what we are calling a socio-technical innovation system builder; and second, it has enabled the building of local capabilities at this cosmopolitan level.

We can see, then, that these projects were important for niche-building in Kenya. They were initiated primarily to test technologies but generated significant effects beyond the first-order learning that SNM would expect of such technology-testing experiments, essential though this first-order learning is to creating the detail of visions. The projects also generated second-order learning for actors within and outside the project networks, and this resulted in altered expectations and changes to behaviour. We cannot be certain that the learning and other effects would not have happened without the projects, but we can see that the private sector would have considered such experimentation risky. Donor-funding gave some protection against these risks, and the experiments provided a means to test assumptions as well as technologies.

But the experiments also brought local actors together in a way that enabled rich interactions over many years, thereby facilitating the exchange of information, and the collectivising of expectations and visions. We also saw that EAA was central to much of the activity discussed here (indeed, they have been central to much activity not examined in this section), and this helped them to become an increasingly skilful cosmopolitan actor (or socio-technical innovation system builder). They identified project opportunities, attracted funding, managed projects and networks of actors, accumulated knowledge and built local capabilities at the cosmopolitan level.

By contrast, the pico-solar experience was one in which the networks were fragmented, expectations were not widely collectivised – indeed, they were contested – and learning was, for the most part, individual rather than collective. Indeed, because learning was poorly articulated, it was difficult to form expectations that *could* have been collectivised. If expectations had been collectivised, then it might have increased the chances of attracting other actors and resources to experiments that could generate further learning.

Market surveys

As we might expect, the various market surveys provided a large amount of detailed information about both the demand and supply sides of the Kenyan PV market. In SNM terms, we can characterise this as predominantly first-order learning – that is, generating finer detail about what is already generally understood. However, it is important to recognise that the surveys occasionally generated information that challenged the assumptions of different actors; that is, we can identify some second-order learning.

While the Hankins (1990) and Acker and Kammen (1996) surveys provided some useful information that helped to detail both supply-side and demand-side practices, they were based on very small samples. The ESMAP-funded survey of 410 households was much more significant. It generated a great deal of first-order learning that enabled a much finer articulation of the market (in the descriptive sense), particularly the demand side. On the basis of this articulation, it was possible to express a persuasive socio-technical vision of PV in Kenya; the objective of rural-household demand for basic electrical services was being provided through the means of PV systems sold in a private market. Further, the observation that the market was moving to smaller systems suggested an extension to this vision or, in some ways, a new expectation: access to electrical services could be deepened to include poorer groups among the population by introducing more 'pico-electricity' products onto the market and providing finance packages 'to lower the initial cost' (Hankins *et al.* 1997, p. 52).

Indeed, EAA had already shown an interest in pico-electricity products, having test-marketed solar lanterns. Notwithstanding this experiment with the lanterns, the precise details of these pico-electricity products were not yet defined; neither were the details of the finance packages recommended in the survey report (Hankins *et al.* 1997, pp. 3, 52). Both these aspects of the expectation were the focus of projects that got underway almost immediately – so quickly,[24] in fact, that EAA had probably formed the expectation prior to the survey, making use of the results to help them collectivise it. The Jua Tosha and BatPack projects went some way to articulating the details of pico-electricity products, while the process of articulating finance packages was the focus of another project (see the discussion of an EAA-managed ESMAP project that experimented with several different approaches: Hankins and van der Plas 2000, and see Byrne 2011 and Rolffs *et al.* 2015 for an analysis of these). This expectation persisted over time and was adopted – perhaps adapted, in conjunction with experiences from elsewhere, by many actors in the development regime, and in PV niches in Kenya and other developing countries.

And we have seen the development regime fund projects that have served to articulate it – envision it, so to speak – while trying to realise its promise, particularly with regard to consumer credit, as micro-finance has emerged as a favoured development tool.

However, Jacobson's research provided a refinement to this expectation, perhaps even a challenge to some aspects of it. His findings concerning intra-household energy allocations refined part of what had become a highly collectivised vision – the benefits in the household of PV-powered light compared to kerosene, especially for women and children. It is perhaps too early to assess whether this will cause any second-order quality change in expectations or visions among actors in the development regime or PV niche, but the dominance of this vision seems to be intact for now. Jacobson's other challenge was that extending credit would not extend access to PV-generated electrical services. As mentioned above, there is currently a great deal of interest in the use of micro-finance to extend services to improve the lives of poorer groups in developing countries, and this continues to be tested together with PV systems. However, there are signs that this is changing. Hankins has begun to talk of 'smart subsidies', arguing that the PV markets have grown quickly in industrialised countries because of generous subsidies. And the GEF has introduced a form of smart subsidy into TEDAP, a recent World Bank electrification project in Tanzania. We are not suggesting that this is because of Jacobson's research, merely that it may have been part of this move away from the rhetoric of 'pure' market forces. More recently still, consumer finance is being tried in Kenya and other countries through the introduction of pay-as-you-go payments made using mobile phones (see Rolffs *et al.* 2015).

The STEP survey also, to some extent, challenged an aspect of the dominant socio-technical vision of PV in Kenya. Although little work had been done to study solar technicians in the market, there was an assumption that they were earning a living by installing and maintaining systems. Jacobson challenged this by showing that the majority of technicians could only secure an occasional job in the PV sector, and so it was just one of many sources of income. The survey also achieved three other things. First, it showed the extent of 'coverage' of technicians in the country and codified these findings. Second, by employing two solar technicians to administer the survey, it helped to connect technicians across the country together in a way that had not been attempted previously. An interesting outcome of this was that the technicians created their own association, KESTA, as a way to promote their interests. In doing so, they created a channel for collectivising an expectation that might express their perspective within the Kenyan PV sector. So the survey stimulated a network effect. Third, the STEP survey appears to have contributed to developing a different, but standardised, training package for technicians.

Perhaps the most interesting part of the ESD (2003) report, from our perspective, is that it makes a list of detailed recommendations that express an accumulation of knowledge gained by EAA/ESD over the preceding decade. Moreover, the recommendations can be read altogether as a clear and finely articulated socio-technical vision of how PV diffusion through Eastern African markets can be successfully achieved.

The recommendations also indicate a slight departure from the more full-blooded free-market approach to PV diffusion of the earlier reports. A notable addition here is the advocacy of smart subsidies, based on the argument that subsidies have been important for the growth of PV markets in industrialised countries. This certainly marks a change of assumptions, and we could interpret this change as a *vicarious* second-order learning effect, as PV markets in industrialised countries provided the source of learning. But it may also reflect the shift in thinking within the development regime – the Post-Washington Consensus – and the 'rediscovery' of the role of government. Whether this was the case or not, a significant part of the explanation for the interest in subsidies may lie in the desire to raise quality in the PV market. In this sense, it recognises a market failure: the Kenyan PV market, applauded for being 'undistorted' by subsidies, has seen a downward spiralling of quality as competitive pressures have caused private actors to cut costs wherever possible; smart subsidies are seen as a way to add value to better quality systems so that private actors are encouraged to eschew the race to the bottom.

Apart from this shift towards the use of subsidies, however, the document could be seen as a statement of the knowledge of the PV sector in Kenya that had been cultivated by a number of actors over the course of at least a decade – longer in the case of Hankins. It expresses a clear vision of how to diffuse PV systems and, because it came from ESD – a well-recognised 'cosmopolitan actor' in the region at the time (now Camco Advisory Services) – it carried authority and could be interpreted as the *dominant* socio-technical vision within the PV niche.

Conclusion

This chapter has charted the arrival of PV systems into East Africa via donor-funded community services projects that later fed into the emergence of a household market in Kenya. We have also looked in detail at how the household market potential began to be exploited and how the idea was picked up by other companies. The analysis has highlighted a range of insights via the application of a socio-technical approach using the analytic categories of SNM. It also makes it clear that a small number of key actors played important roles in disseminating knowledge about the emerging PV niche and recruiting others to a broadening network. In Chapter 7, we look in more detail at the implications of these observations in the context of our broader concern with Socio-Technical Innovation System Building.

In Chapter 5, we turn to analysing how, in the early 2000s, PV niche actors began to interact with the policy regime in an attempt to influence Kenya's new energy policy, with mixed results. We also describe and analyse two significant recent occurrences. First, there is now a rapidly emerging market for pico-solar lanterns, created through the actions of the IFC-funded Lighting Africa programme and exploited by a growing number of private sector actors. And, second, the first solar module assembly plant in Kenya has now been in operation for several years, the result of a joint venture between a European company and one of Kenya's largest suppliers of PV modules and batteries.

Notes

1 The programme provides vaccines against BCG, DPT (diphtheria, pertussis, or whooping cough, and tetanus), polio and measles (Henderson 1989, p. 46, Figure 1).

2 'Shamba' is a Swahili word that can be translated to mean 'farm', although it can be used for anything from a plantation to a small plot of cultivated land, and it also has connotations of 'rural' (Johnson 1939, p. 416).

3 'Kidogo' is a Swahili word that can translate as 'small' (Johnson 1939, p. 76, see the entry '-dogo').

4 'Harambee' is used in Kenya to mean 'self-help', and is the national motto (Barkan 1994, p. 19).

Harambee, or self-help, is a pervasive movement that has become a major arena of rural politics and has shaped the structure of peasant-state relations in that country. With its fifteen to twenty thousand community development organizations scattered across rural Kenya, this self-help movement engages just about all rural dwellers, most politicians and many state personnel. The primary activity of these organizations is the construction of social-service infrastructure by the residents of rural communities in order to meet their locally defined needs ... nursery, primary, and secondary schools, village polytechnics, cattle dips, health centres, water projects, etc. (Barkan and Holmquist 1989, pp. 359–360).

5 Although Hankins describes this as a charge regulator, it is likely that the device was actually a change indicator, as Burris used such self-designed indicators in later installations (Hankins 1990; 1993, p. 35).

6 The 'coffee boom' is actually said to have occurred during the period 1975/1976 to 1978/1979 (Akiyama 1987, pp. 6, 8; Bevan et al. 1990, p. 359). There was a peak in the value of coffee exports in 1977, following which the value fell back below USD 300 million as of 1980, remained quite steady, and then peaked at a similar level to the 1977 value in 1986. Bevan et al. (1990, p. 359, citing an earlier study of theirs, Bevan et al. 1987) state that coffee producers in Kenya, unlike those in other coffee-exporting countries at the time, received significant earnings from the boom because 'export taxes were negligible'.

7 It is not clear what an independent placement is, exactly. It may be that Peace Corps volunteers were placed in organisations to help build capacity in those organisations. In the case of Hankins' work with Burris, this would not be the arrangement and so it may have been considered 'independent'.

8 The payback time was about two months and there were an estimated 125,000 stoves sold by the middle of 1985 (Jones 1986, p. 18).

9 Hankins was not present when the Governors visited Burris' system (Hankins 2007, interview).

10 A diode is connected in series between the module and battery to prevent discharge from the battery when the module voltage is lower than the battery voltage, as would happen in darkness. Muchiri (2008, interview) says that modules with attached diodes attracted duties while those without did not. In order to avoid duties, the international suppliers would be asked to send diodes separately.

11 Although Total were interested in selling solar equipment in Kenya (and perhaps elsewhere), Rioba characterised their motivation as a public relations exercise; that is, it was more of an attempt to look environmentally responsible than a serious attempt to develop sustainable technology markets (Rioba 2008, interview). Nevertheless, before they started selling PV, they had about 70 per cent of the solar water heater systems market in Kenya (Hankins 1990, p. 67).

12 SolarNet was a network of actors interested in the promotion of renewable energy technologies.

13 ESMAP was run out of the World Bank and had the aim of helping developing countries make energy assessments and development plans.

14 APSO was the Agency for Personal Services Overseas; a now-defunct agency of the Irish Foreign Office that was focussed on development assistance through quasi-voluntary work.

15 These 'systems' consisted of low quality PV components bought piecemeal and assembled without any design considerations.

16 In practice, the battery was measured to have a 30 Ah capacity (Ochieng et al. 1999, p. 22).

17 EAA (2001, p. 4, n. 5) states that the battery probably had a lower capacity than the manufacturer claimed.

18 Rodson was the local company that designed and assembled the charge controller.

19 The information provided to the focus groups was in the form of pictures of the product concepts, and it proposed functional and technical specifications (Osawa 2000).

20 DIY Solar was an idea developed by Graham Knight in Ashford, Kent, in the UK. He made use of 'discarded' amorphous PV modules from Intersolar, which he cut into smaller pieces and to the back of which he soldered wires. He was then able to power devices such as radios (Blyth 2008, interview).

21 In fact, this Musinga et al. (1997) survey of 1000 households may not be focussed on the PV niche; it may have focussed on non-PV households.

22 ESD was a later incarnation of Hankins' company EAA, after it was bought in the late 1990s by the UK company Energy for Sustainable Development.

23 PVMTI was an International Finance Corporation project that made USD 5 million available for finance in the Kenyan PV market, responding to a perceived finance barrier that was preventing the market from expanding. Beginning in 1998, it intended to lend to both suppliers and MFIs to help reduce the price of PV modules to consumers, thereby releasing pent-up demand and transforming the market. Along with Paula Rolffs, we analyse the PVMTI intervention, and more recent developments in micro-finance for PV, elsewhere (see Rolffs et al. 2015), and we also provide a brief analysis in Chapter 6.

24 The Jua Tosha and BatPack projects got underway in 1997, before the household survey report was finalised (Ochieng et al. 1999, p. 9; EAA 2001, p. 5). The finance project was already in a preparatory phase in December 1996, as the household survey was beginning (Hankins and van der Plas 2000, p. 25, Box 5–1).

5

POLICY REGIME INTERACTIONS AND EMERGING MARKETS

Introduction

In Chapter 4, we began our innovation history of solar PV in Kenya by charting the emergence of PV in the country and the eventual articulation and development of a household market. In this chapter, we present the second part of our innovation history. We begin by analysing how PV niche actors began directly interacting with the policy regime in attempts to influence Kenya's new energy policy. This had mixed results and reflects an uneasy relationship between PV niche actors and some influential figures in the policy regime – an uneasy relationship that continues to reverberate through policies that affect niche development.

We then move on to analyse two significant recent developments. These include the emergence of a market for pico-solar lanterns – operating in parallel with that for SHSs – which is drawing in new actors and offering new hope of addressing the needs of poorer groups. This is a particularly significant development as the creation of the pico-solar market was in large part driven by the deliberate interventions of the International Finance Corporation (IFC) funded initiative, Lighting Africa. The Lighting Africa activities resemble many aspects of the Socio-Technical Innovation System Building interventions we articulated in Chapters 2 and 3, as opposed to any kind of Hardware Financing or Private Sector Entrepreneurship approach. They certainly sit in stark contrast to the Photovoltaic Market Transformation Initiative (PVMTI) – an earlier IFC-funded intervention we mentioned in Chapter 4, and which we analyse briefly in Chapter 6.

Another development in the Kenyan solar PV sector is an investment in the country's first-ever solar module assembly plant, which we see as signalling other possibilities related to our discussions in Chapter 2 and Chapter 3. In this case, the significance arises from understanding the extent to which the development of consumer markets for sustainable energy technologies might subsequently foster the development of indigenous industrial capabilities and capacity. If such a 'sequence' of industrial development is in any useful sense generalizable to other contexts, then our case hints at answering concerns about the desirability of employing low carbon development or green growth strategies in developing countries. The Innovation Studies literature – as we explained in Chapter 2 – has certainly identified sequencing of capability development in what are now the emerging economies. We can only speculate at this time about whether the solar assembly plant is a sign of something similar to come in Kenya, but further research might yield useful insights.

Let us begin, then, by looking at the interactions between the solar PV niche and the policy regime in Kenya.

Policy regime interactions

By the end of the 1990s and the beginning of the 2000s, PV niche actors had begun to interact more significantly with some in the Kenyan policy regime. The results of these interactions were mixed, not least because some policy actors refused to accept PV as a means to increase electricity access in rural areas. This section discusses several of these interactions and reveals how, at this policy level, they generate much more obvious political conflicts than those at the niche level. We begin with the process of formulating PV standards, which was the first of these more substantial interactions. This was followed soon after by a concerted effort to directly influence Kenya's second energy policy, which was published in 2004. While there seems to have been some success in influencing this policy, more recent attempts have seen niche actors marginalised. But there has been more success in formulating PV regulations, at least in the minds of some in the PV niche. We discuss this process before finishing with a short consideration of the regime of taxes and import duties on PV in Kenya.

PV standards

For many years, the lack of standards for PV in Kenya was a recurrent issue, raised time and again during workshops, seminars, and in the writings about the market (Hankins 1990; 1993; Hankins and Bess 1994; Acker and Kammen 1996). The donor-funded installations, at least those such as the World Health Organisation-Expanded Programme on Immunization (WHO-EPI) systems mentioned in Chapter 4, had their own standards but there were no Kenyan standards that could be applied in the private market. We have already discussed the attitudes to technical standards of some of the pioneers in the Kenyan PV market, and some of the practices on both the supply and demand sides that emerged from the highly competitive environment of the late 1980s. Despite the many calls for technical standards that resulted from the recognition of these practices, it was not until the mid-1990s that there appears to have been any attempt to persuade the Kenya Bureau of Standards (KEBS) to do something about the issue. This initial attempt to get KEBS to formulate technical standards failed (Gisore 2002, p. 47), possibly because it was attempted only by a single actor (Loh 2007, interview). However, in 1998, KEBS 'revisited' the development of PV standards following 'increased demands from various quarters' (Gisore 2002, p. 47). It is not clear what this means, but there was mounting evidence that there were serious problems with some of the products and practices in the PV market (Hankins 1990; Hankins and Bess 1994; Acker and Kammen 1996; Hankins et al. 1997), and there were moves to develop international PV standards (PVGAP 1996; 1998).

In any case, KEBS decided to initiate a standards process for all the renewable energy technologies, starting with PV. This got underway officially on 28 April 1999, consisting of a committee of about 12 invited stakeholders from the renewable energies sector in Kenya (Gisore 2002, pp. 47–48; Loh 2007, interview). The process of writing the standards consisted, in essence, of monthly meetings for which the committee members reviewed draft standards such as PVGAP, wrote outlines, and discussed what should be included, excluded, and what needed work (Loh 2007, interview).

While this sounds like an essentially technocratic process, there is some evidence that it was not straightforward. Two extracts from a presentation given by Gisore, the KEBS representative on the committee, hint at the sometimes contentious deliberations that unfolded and why they were so (Gisore 2002, p. 49):

> As an activity that touches on the social and economic aspects of stakeholders, agreements on the standards and codes of practice have been based on consensus. This has not been easy. Many times members have had to vigorously demonstrate the negative or positive effects which certain requirements in the standards or codes of practice will have on the subject matter. Many times consensus has not been reached in a single sitting, and this accounts for the fact that it has taken almost three years to have most of the standards get approval as Kenya standards.

And, Gisore (2002, p. 50) states: 'For those components manufactured locally, due considerations were given to ensure that the standards did not serve to push [their manufacturers] against the wall.' Nevertheless, by the time Gisore gave this presentation in August 2002, much of the standards work for PV was complete, even if not everything had been formally agreed (Gisore 2002).

Some time during 2001 to 2002, the committee members began to discuss the idea of forming an association. The argument, according to Loh (2007, interview), was that it would be 'better that the association has its rules and governs itself before the Government comes in and puts its hand into saying all these things and getting licenses. Better a well-regulated industry …'. Both KEBS and the Ministry of Energy (MOE) were 'very keen' on the idea and the Kenya Renewable Energy Association (KEREA) was 'very quickly registered, in August 2002' (Loh 2007, interview). One of the first efforts of KEREA was to conduct a technical evaluation of the amorphous silicon modules on the market in Kenya.

The 'amorphous question' (Ochieng 1999, p. 19) was something of a refrain in Kenyan PV circles, and there had been a major study of the performance of the modules available on the local market, conducted in 1999 by EAA, RAEL at the University of California and STEP at Princeton University (Duke *et al.* 2000). That study found that one manufacturer's amorphous modules performed very poorly, and the company responded by improving its manufacturing process (Jacobson and Kammen 2005, p. 1). Despite this success, new low-quality brands of amorphous modules appeared on the market, and so Arne Jacobson offered to conduct a fresh set of tests for KEREA (Jacobson and Kammen 2005, p. 1; Loh 2007, interview).

There was difficulty in agreeing on the terms of this evaluation but, eventually, KEREA members agreed to the methodology, and a sample of modules was shipped to the USA in 2004, where Jacobson and colleagues performed the tests over the period from September 2004 to March 2005 (Jacobson and Kammen 2005; Loh 2007, interview). Two brands of modules were found to be severely over-rated and so, in line with the terms of the evaluation, the importers of these agreed to remove them from the market (Jacobson and Kammen 2005; Loh 2007, interview). By February 2005, before the results of the module evaluation were ready, KEREA had a code of conduct in place (KEREA 2005). The efficacy of the code of conduct was, therefore, tested almost immediately. According to Loh (2007, interview), it was the peer pressure that KEREA members could bring to bear, based on the agreed code of conduct, that achieved the removal of the sub-standard modules from the market and 'many people were quite chuffed about it that we [KEREA] managed to do something like that ... KEREA became something more credible'.

Despite the apparent success of this initial KEREA effort to remove over-rated modules from the market, there were recurring problems with poor quality products over subsequent years. Even though standards were in place, they were not successful in dealing with these quality issues. This was due partly to weak capacity at KEBS to test equipment (Mboa 2013, interview), poor product quality information and many new products entering the market (one interviewee suggested that there had been dumping of products), for which standards did not necessarily exist. As well as the problems with poor quality products, there continued to be issues with over-selling by vendors and poor installation by technicians. The introduction of a capacity-building component in PVMTI (see Chapter 6) was intended to deal with these issues – vendor and technician training and information for consumers – but it appears that this was insufficient, even if PV actors considered it a step in the right direction. KEBS revisited the standards around 2008, about the same time as the recently established Energy Regulatory Commission (ERC) (see the next section on energy policy) started to develop PV regulations. By this time, it seems that some PV actors were much more willing to support regulation – not just standards – because they were fearful that the PV market was being seriously undermined by the issue of poor quality products and practices. We pick this story up after the next section, which recounts the development of energy policy beginning in the early 2000s and from which the ERC was created.

Energy policy proposals and politics

Around the middle of 2001, the process of preparing a new energy policy began in the Ministry of Energy, and discussions were initiated involving various government departments and representatives from parastatals (Theuri 2008, interview). Except in their individual capacity, no other energy sector stakeholders were invited to participate at this point. However, there was at least some interaction between the ministry and others in the renewable energies private sector. Daniel Theuri, the Acting Director of the Department for Renewable Energy, worked with both Mark Hankins and Bernard Osawa of EAA within the IGAD Regional Household Energy Project, writing a handful of papers related to energy in Kenya (Theuri and Hankins 2000; Theuri and Osawa 2001; Osawa and Theuri 2001), possibly the first substantive and formal collaboration between the ministry and actors in the non-commercial renewable energy sector[1] in Kenya.

Soon after this official MOE process got underway towards the end of 2001, EAA began talking with the UK Department for International Development (DFID) about the possibility of funding an energy policy discussion process. Early in 2002, DFID agreed to fund what became known as the 'Policy Dialogue', and the first session took place on 21 May in Nairobi (Bess 2002, p. 1; Mutimba 2007, interview). Another five meetings took place that year: one each in June, August, September, October and December (Mutimba 2002a, 2002b, 2002c).

The MOE policy was 'already taking shape' by December 2001 (Theuri 2008, interview), but some time in 2002 the UN Development Programme (UNDP) country office was asked to support the process (UNEP 2006). Theuri (2008, interview) states that the draft policy was ready by April 2004, but Mutimba (2007, interview) claims that the Policy Dialogue had managed to get hold of a copy of the draft during 2003, following which they drafted an alternative policy and submitted this to the MOE by the end of the year. Whatever the precise details of the timeline, during which there seems to have been some tension and politics between the Policy Dialogue and the MOE processes (Mutimba 2007, interview), a Sessional Paper was indeed passed towards the end of 2004. However, it took another two years before this became the Energy Act.

During those two years, there were more 'cat and mouse games' between the MOE, parliamentarians and the Policy Dialogue (represented by ESD[2]), as well as interventions by the 'traditional' energy actors, such as the utility companies and those in the petroleum sector (Mutimba 2007, interview; Otieno 2007, interview). In terms of the MOE-Policy Dialogue interactions, one account has it that the MOE 'took the [Policy Dialogue] document and oppressed it a bit' (Mutimba 2007, interview), but used much of it as the official energy policy, while another account claims that the influence of the Policy Dialogue was only really on the charcoal policy (Theuri 2008, interview). It is not possible to verify either of these accounts, but we do have detailed information from Otieno (2007, interview) on how the MOE attempted to have its version of the policy endorsed by the parliamentary committee on energy. For reasons that are unclear, Otieno and Mutimba were present as observers[3] at this meeting. According to Otieno, he and Mutimba realised that the policy the MOE was presenting had 'everything to do with renewable energy extracted', and informed the committee of this. There then ensued the 'cat and mouse games' between the ministry, the committee and ESD. In essence, the parliamentarians insisted that the MOE reinstate the renewables passages, having been briefed by ESD and GTZ about the details. Eventually, partly because of the MOE's 'fear' of the parliamentarians[4] (Mutimba 2007, interview), a compromise was reached, whereby the renewables components were, at least, strengthened again (Otieno 2007, interview). As a result of this experience, Otieno 'realised that the parliamentarians have a critical role in formulating policy and have an upper say when it comes to the ministry'. In response to requests from the parliamentarians, GTZ supported the forming of a network – the Parliamentary Network on Renewable Energy and Climate Change – in which ESDA and others conducted seminars for the parliamentarians on renewable energies (Otieno 2007, interview).

The Energy Act of 2006 is not specific about the nature of the various intentions it states for renewables, but there were practical implications, including a very large project to install PV systems in schools and health centres (Onyango 2007, interview). However, the initiation of this Institutional PV Systems Programme was not due to the Energy Act; it actually began before the Sessional Paper on energy received assent in Parliament, which was a result, it seems, of presidential pressure following an electoral promise to electrify North Eastern Province (Mutimba 2007, interview). According to Mutimba, the MOE decided to go with PV to electrify schools, despite a long-standing resistance within the ministry to renewable energies, because there was no other way to realise quickly the promises that the president had made during his election campaign. Onyango (2007, interview) tells this slightly differently, claiming that the Permanent Secretary (PS) of the MOE was 'the champion' within the ministry for the Institutional PV Systems Programme. Judging by the views expressed during interviews with some of the actors in Kenya, the former appears to be more likely; the PS was apparently well known for his objections to renewable energies and is said to have expressed his views publicly (Mutimba 2007, interview; Otieno 2007, interview).

Whatever the origins and motivations, the MOE started the programme with some pilot installations in one school (Onyango 2007, interview). There were technical problems with the systems, but these were fixed after the MOE employed a long-standing PV engineer – Kiremu Magambo – to consult on the project. Magambo also ran training sessions on PV systems for others in the MOE in preparation for the expansion of the programme (Onyango 2007, interview). The money to be spent by the government on the programme was a significant injection into the PV sector. Up to the end of financial year 2006/2007, the expected spend would be almost KES 257 million (USD 3.7 million approximately, using KES 70 = USD 1). For the next two years, the budgeted spend was to be KES 335 million (USD 4.8 million). Altogether, this would add about 514 kWp to the installed capacity in Kenya (Mbithi 2007, slides 12–18, authors' calculations). These would be additions of the order of 20–40 per cent of the value of the household market at the time (Onyango 2007, interview; Mutimba 2007, interview; authors' calculations). Given this substantial commitment of resources, it appeared to mark the beginning of what might be a more supportive policy environment for renewables in general, as evidenced by the budgets for energy reported in ROK (2007) and shown in Table 5.1.

However, there were mixed feelings about the Institutional PV Systems Programme. While it was being welcomed as a positive move in general, there was some indications that it had raised the price of PV to the consumer, and there were suspicions of corruption within the procurement process (Mutimba 2007, interview). There were also some issues over who could win contracts; despite an aim to include local technicians and companies in the work (Onyango 2007, interview), in order to get a contract, a tendering company needed to have a 'secure' financial base and this limited participation to a handful of large companies (Rioba 2008, interview). Still, the programme continued and, according to MOE (2013, p. 59), 945 institutions – including primary and secondary schools, dispensaries, health and administrative centres – had PV systems installed by the end of 2012.

A further institutional development created by the Energy Act was the establishment of the Energy Regulatory Commission (ERC). This had the mandate to regulate production, distribution, supply and use of renewable and other forms of energy. Under this mandate, it began the process of developing PV regulations, in consultation with the KEBS sub-committee on PV and wind. As with the standards process that preceded it, there were occasional tensions between different actors on the committee over the stringency of the regulations. For example, Mboa (2013, interview), who was now managing the KEBS committee following the

TABLE 5.1 Expected energy sector outcomes up to 2010

| | Estimate | | | Projected estimates | | | | | |
| | 2007/08 | | | 2008/09 | | | 2009/10 | | |
Programme	KES billion	USD million	%	KES billion	USD million	%	KES billion	USD million	%
Energy sector recovery	6.29	89.80	35.72	3.22	46.02	22.30	3.22	45.94	22.36
Energy efficiency	0.04	0.51	0.20	0.04	0.51	0.25	0.04	0.51	0.25
Rural electrification	5.74	81.97	32.61	5.74	81.97	39.71	5.22	74.62	36.31
Renewable development	3.47	49.54	19.71	3.97	56.78	27.51	4.48	63.96	31.13
Fossil fuel development	2.07	29.56	11.76	1.48	21.13	10.24	1.43	20.45	9.95
Sub-total	17.60	251.38	100.00	14.45	206.41	100.00	14.38	205.49	100.00

Source: Adapted from ROK (2007, p. 26, Table 4–0-0).

secondment of Gisore to the African Regional Standards Organisation, describes the discussions over warranty periods for various PV components.[5] The government and consumer representatives on the KEBS committee wanted lengthy warranties, but the private sector representatives were unhappy with this, claiming that it could put them out of business. For some of the components, the private sector actors claimed, the warranties demanded in the draft regulations exceeded those given by the manufacturers of those components. Eventually, the chair of the committee, Kiremu Magambo offered a compromise that was acceptable, and the regulations were eventually gazetted in late 2012. Before discussing some of the other details of these regulations, it is worth noting recent developments in energy policy.

Following the new constitution in Kenya, enacted in 2010, and a confluence of other factors, there is a need to formulate an updated energy policy. Perhaps chief among these other factors is pressure to promote much faster economic growth (Kenya has the goal of becoming a middle-income country by 2030), which is being hampered by a combination of high prices and unreliable supply of grid-based electricity (Newell et al. 2014). Of less concern to some in the Kenyan Government is the need to promote development that is climate-compatible, but there are those who see opportunities in steering Kenya along such a pathway and, in many ways, renewable energy-based electricity generation could help deal with the problems of high prices and unreliable supply. Furthermore, many donors, who have some influence over Kenyan energy policy, have been pressing for a low carbon development agenda. This agenda has started to pique interest in the Finance Ministry, mainly because of the possible flow of significant resources from climate finance.

In alignment with these drivers, the Kenyan Government began to introduce feed-in tariffs (FITs) in 2008 for a range of renewable energy technologies. Solar did not feature in the early days of the FIT but does so in later iterations, including in the latest update from December 2012. However, the smallest off-grid PV installation project eligible for the FIT subsidy is 500 kW, much larger than any SHS found in Kenya (MOE 2012, p. 16). So there appears to be a continuing lack of interest by the Kenyan Government in regard to SHSs. And this is underscored by the brief (three-page) entry PV has in the latest draft of the new energy policy (MOE 2013, pp. 58–61) – an entry that also includes reference to solar water heaters. While there is acknowledgement in the policy that Kenya has one of the most successful off-grid SHS markets in the developing world, there are only vague goals for promoting the systems further. The most specific intentions for PV relate to the institutional programme discussed above and to the conversion of a number of large remote diesel installations to diesel-PV hybrid systems.

It is likely that the form of this new energy policy reflects the continuing exclusion of actors who are working in the small-scale PV sector. According to Newell et al. (2014), no such actors were invited to consultations during the drafting of the policy, and attempts by KEREA to provide inputs to the process were ignored. Furthermore, it seems, the former PS of the MOE[6] continues to wield power over the energy sector, and it is alleged that he actively intervened to undermine any support for PV. He is said to have ensured that the FIT, for example, was set low for solar so as to make it unattractive to investors. It is not yet clear whether the new PS has a more favourable view of PV, but it is clear that some renewable energy technologies are seen as attractive. Most notably, geothermal is stimulating enormous interest from many in the energy sector: government, private sector actors and donors. Among the private sector actors are large businesses from a range of sectors, including manufacturing. They are, of course, interested in low electricity prices and reliable grid-supply. For them, geothermal offers the possibility of meeting their needs at scale. Donors appear to be interested because geothermal is low carbon and so aligns with their agendas for climate-compatible development. The government appears to be interested because geothermal could provide a way to relieve pressure from the powerful manufacturing lobby (on prices and grid reliability) and from grid-connected consumers. And the large capacity increases that geothermal could realise would underpin the increased economic growth that Kenya needs in order to achieve its goal of becoming a middle-income country.

Although recent policy in favour of low carbon development in Kenya might have been expected to benefit the promotion of SHSs and SPLs, it is clear that the government has prioritised least-cost economic growth over other development aims, such as energy access. Any aspirations for increasing energy access in off-grid areas seem to be resting on a hope that the small-scale PV sector will continue to operate as a private market for which the government simply sets the rules that the regulator enforces. There has certainly been an increased effort to regulate the PV market in recent years, as we shall see in the next section. However, there are also dangers in assuming that the market can be successfully regulated, even though many of the actors involved are now relatively enthusiastic about this approach. As we shall see, the standards and regulatory infrastructure in Kenya is lagging behind the development of the market in both capabilities and capacity.

PV regulations

As discussed above, the ERC was established in 2006 and subsequently mandated to develop regulations for the PV sector. The Kenyan PV market was suffering persistent problems with poor quality products, and the quality of installations and after-sales service were also low. Although there were PV standards in place, they were not being enforced. Altogether, these quality issues were perceived by the bigger players in the Kenyan market as a problem for them and for the market as a whole. As a result, they urged the government to introduce regulations (Mabonga 2013, interview). The regulations were developed in consultation with PV actors – particularly using the PV standards sub-committee managed by KEBS – and eventually came into force in September 2012.

Among its many regulations, there are requirements for all those involved in some way in PV to be licensed: manufacturers, importers, suppliers, vendors, contractors and technicians (ERC 2012). Various classes of licence cover these different groups but they all require renewal annually along with – in most cases – payment of a fee. In the case of technicians, for example, there are three levels of competence recognised: T1 (basic), T2 (intermediate) and T3 (advanced). T1 technicians, it appears, do not need to pay for a licence, but they must have achieved a minimum level of approved training and two years of PV installation experience. Indeed, all classes of technician are required to meet approved training and experience criteria. However, T2 and T3 technicians have to pay KES 2500 (USD 24) and 3750 (USD 37) respectively, covering application and first licence. Renewal then costs KES 750 (USD 7) and 1000 (USD 10) respectively each year. Considering the numbers of technicians and vendors already active in the PV sector across Kenya, and that it is an offence under the regulations to carry out these activities without a licence, there is the potential for an enormous administrative burden on the ERC (the licensing authority) at the very least. And yet, the RE department has only four or five staff (Mboa 2013, interview). The register of licensed technicians as of November 2015 had 300 names (ERC 2015b), up from 37 in January 2014 (ERC 2014b) and the contractors register (which includes all other categories) had 229 as of November 2015 (ERC 2015a), up from just three in January 2014 (ERC 2014a). While there has clearly been substantial progress in issuing licences, it is not clear whether it is enough. Ondraczek (2013, p. 409), for example, reports that there may be upwards of 2000 technicians serving the market. Of course, given Kenya's history of untrained technicians, we might expect that not all of these 2000 will be able to get licences. Nevertheless, the disparity in numbers between those on the register and those who might be working in the field is suggestive of what could be an unmanageable administrative burden for the ERC.

The requirement for approved training has meant the need to develop a nationally-recognised PV syllabus for the three classes of technician. Initiated by KEREA (Anonymous 2013a), and supported by UNDP, this process got underway in July 2012 with a five-day workshop, to begin developing the curricula. A range of actors was present at the workshop, where the details of the syllabus and tests for each of the classes of technician were discussed and drafted (KEREA 2012). Jomo Kenyatta University of Agriculture and Technology (JKUAT) led the subsequent development of the curricula, in collaboration with ERC and funded by the Japanese International Cooperation Agency (JICA) (Mabonga 2013, interview). Chloride Exide – a battery manufacturer and one of the largest PV distributors in Kenya – were asked to assist with research into appropriate PV system components and to provide sizing of a range of systems as well as to install them at various field locations. The research, syllabus content development, sizing and installations were done in the period up to November 2012. After this, Paul Mabonga of Chloride Exide visited the installations to conduct monitoring of the systems up until February 2013. From March 2013, the curricula were ready to be sold to colleges in Kenya.

The syllabus provides for a one-month course module to be incorporated into a standard electrical technician training course, and was piloted with a group of 100 technicians before being offered for purchase (Mabonga 2013, interview). Fees for the training are expected to be about KES 15,000 (USD 146). Those technicians who have already accumulated experience have the option of taking a test instead of the training and would pay about KES 5000 (USD 49). The first formal courses were expected to run from September 2013. The National Industrial Training Authority (NITA) has equipped four training centres to run the module and test technicians, and KEREA is said to be trying to equip another ten (Anonymous 2013a). However, there seems to be a lack of resources to suitably equip these centres, and so it is not clear whether the technician training will be able to meet what could be an overwhelming demand.

In parallel with the development of regulations for the PV sector, there has been a process of developing standards for SPLs. However, this has been led by Lighting Africa rather than KEBS. Again, the process was motivated by the experiences in the market with many poor-quality products; in this case, the products were solar lanterns but the wider experience with poor-quality SHSs was having an influence on the perception of SPLs too. Indeed, the perception of poor-quality SPLs was well founded, as Lighting Africa discovered when they tested a range of 14 lights in 2009 and only one passed (Anonymous 2013b). Following this, Lighting Africa initiated research to find out what minimum standard of quality would be acceptable to lantern-users. They then worked with the global lighting industry, advocating for better-quality lanterns and awarding prizes for the best products at conferences in 2010 (Lighting Africa 2011) and 2012 (Lighting Africa 2013a). When Lighting Africa tested another range of 20 products in May 2010, they found eight passed the minimum standard (Anonymous 2013b).

Lighting Africa further pursued the development of these standards and managed, after two and a half years, to get them adopted by the International Electro-technical Commission (IEC) (Anonymous 2013b). Once adopted by the IEC, they can be adopted at national level and, at the time of the research in 2013, KEBS were considering this (Mboa 2013, interview). Alongside this, Lighting Africa has been working with the University of Nairobi to establish a testing laboratory to help in the enforcement of the standard, and to provide a place where products can be screened by importers before they commit to buying large stocks (Anonymous 2013b). The test facility could also act as an education and training tool, enabling students to gain experience with solar testing procedures and equipment. However, before the laboratory can enforce the standards, it must itself be accredited, which means it has to comply with an ISO standard itself. This, too, was in process at the time of the research, with assistance from KEBS (Mboa 2013, interview).

Although there was progress in regard to standards for SPLs, there was also some disquiet about the test procedures, their cost, their value, and their stringency. Some private sector interviewees, for example, were not entirely happy with the quality assurance offered through Lighting Africa and claimed that the programme was promoting sub-standard products alongside those approved. Other issues related to the number of different types of lanterns available on the market (around 40–45), with some actors saying that there were too many, causing confusion among custo-mers; that customers were not buying approved products, and so there was ques-tionable value in paying the USD 6000 to go through the quality assurance test; and one interviewee claimed that Lighting Africa was not interested in quality products, suggesting that it was really only interested in sales figures and market growth. Whether these contentions are indications of what would be expected in a nascent market or are more fundamental problems remains to be seen. It is clear, however, that Lighting Africa's attempts to develop and enforce standards are in step with the wishes of both the major small-scale PV sector actors and those in the regulatory regime. In any case, it appears that the standards were officially approved in June 2014 (KEBS 2014, p. 11) and gazetted in October (ROK 2014). Alongside this, the University of Nairobi Lighting Laboratory achieved accreditation for screening of solar lanterns, although full testing takes place in either Germany or the USA (UON 2014).

Taxes and duties

The issue of VAT and import duties on PV and associated equipment has been an abiding feature of the Kenyan PV niche for decades. We have already discussed the issue in Chapter 4, where several observers of the Kenyan PV niche could not agree whether import duty and VAT removal in 1986 had been passed to the consumer. Since then, taxes and duties have been applied and removed many times and at different rates, and on different parts of PV systems. For example, Jacobson (2004, p. 143, Table 16) shows cumulative tax and duty rates rising steeply on PV modules in 1992 before falling back in steps until they are zero-rated again in 2002. Since 2002, there has tended to be a more favourable duty and VAT climate in Kenya for PV, perhaps explained by the close relationship with some parts of government enjoyed by members of KEREA while working on standards (Newell et al. 2014). While any rises in taxes and duties tend to cause alarm in the PV niche, the market has continued to grow (Ondraczek 2013). This is not entirely surprising, given that most SHSs are bought by the middle class, even if higher prices do hurt them.

More recent moves on tax, however, look likely to be damaging to some in the PV niche. As discussed below, Kenya has seen rapid growth in a pico-solar market since about 2009. The customers in this market are much poorer than those who might buy SHSs, but the government imposed 16 per cent VAT on solar goods from 1 October 2013, and many of the private sector actors in this pico-solar market whom we interviewed said they had seen drastic falls in sales as a result. As Newell et al. (2014) observe, the Kenyan Government currently needs to raise tax revenue, and so those actors who do not wield lobbying power are likely to suffer the burden of this need. With the government prioritising geothermal in its low carbon development plans, off-grid PV – whether SHSs or pico-solar – appears to be of little relevance currently.

Socio-technical analysis of policy regime interactions

PV standards process

The process of formulating PV standards in Kenya was a site for considerable first-order learning, as actors were focussed on the details of what those standards should be. Clearly, this entailed substantial technical discussions that encompassed draft standards such as those being developed through PVGAP, the experiences and expertise of the local niche actors, and the requirements of the Kenyan regulatory regime.

But we can identify some second-order learning that was also important in the process. This second-order learning occurred much earlier for some niche actors when they realised that there were quality problems[7] in the market. Based on this realisation, they formed a new expectation, perhaps even vision, in which the solution to these quality problems was to regulate the market using standards. They made repeated attempts to collectivise their understanding by expressing a vision of a PV market that was successful and of high quality, with the enforcement of standards as the means to achieve this objective. In fact, they presented two visions. The other, which was to some extent being realised in the market, was a negative vision in which consumers were losing, and business would fail, because of bad practices. Eventually, KEBS was recruited to this vision and initiated an official standards-making process, although it is not clear why this second-order learning did not occur sooner for them.

The process also contributed to the enhancement of networks within the niche, as was the case with other projects we have already discussed. For some of the actors involved, their only interactions with others in the niche had been an occasional business deal; now they were meeting regularly to discuss issues other than business (Loh 2007, interview). It was out of this close interaction that they formed an industry association (KEREA). We could see this as a second-order learning experience in that they formed a new expectation, related to the standards issue, in which one of the objectives was a high-quality PV sector[8] that could be achieved by self-regulation of the factors not covered by the technical standards. This expectation was then envisioned to some extent by the formulation of a code of conduct, and the initial embedding of this when they managed to persuade the 'guilty' KEREA members to remove low-quality modules from the market.

One other aspect of the standards process, for which we have only *suggestive* evidence, is the contention generated by this kind of action. We can interpret standards as socio-technical visions: they are highly detailed prescriptions for certain aspects of action and so intended to formally institutionalise particular behaviour. In this sense, the niche actors on the committee were negotiating a vision of serious importance to them; each could be affected in different ways by the outcome of the process – that some could be winners and others losers, depending on the constraints imposed by the institution. Gisore (2002, pp. 49–50) hints that this was indeed how some of the committee deliberations unfolded and is more explicit when he states that the process included consideration of the consequences for local actors. Unfortunately, we cannot examine these negotiations because we do not have the evidence and, therefore, cannot assess to what extent they shaped niche development. But we can recognise that important niche-shaping action resulted from the process, and that the process was inherently political.

Energy policy-making process

Both power and politics had important shaping effects on the niche developments we have discussed in relation to energy policy-making. The Institutional PV Systems Programme was the result of *ad hoc* policy-making realised because of the power of the President's Office, and driven by the raised expectation among voters of electrifying their part of the country. And the official process of preparing the 2004 national energy policy became a political struggle with the unofficial process of the Policy Dialogue. The final outcome of that struggle – the Energy Act 2006 – was a compromise achieved through the exercise of the power of parliamentarians. Of course, these outcomes were not simply the result of power and politics; expectations, learning, networks and institutions – as SNM posits – were all involved as well.

The Institutional PV Systems Programme was initiated because of the expectation of electrification that the President had, it is claimed, collectivised during his election campaign. The only way that the MOE could realise this quickly was with PV systems. However, following years of neglect of renewables by the ministry, their internal capacity was poor. So, the MOE had to employ a niche actor to help them envision the expectation: troubleshoot their first system, design systems, train MOE staff and so on. The impact for the PV niche was significant. While it created some big winners among those who won contracts, it also created some disquiet among other actors. In the case of PVMTI, as we discuss briefly in Chapter 6, disquiet stimulated actors to collectivise a new expectation and to seek a shift in policy to more capacity building. This does not seem to have been the case with the Institutional PV Systems Programme. Perhaps, unlike PVMTI, there were at least some winners in the programme, and this might have fragmented any efforts to collectivise an alternative expectation.

The formal process of preparing policy, as we might anticipate, was a highly *political* activity, more so than the other activities we have studied. The number of interested actors, and the consequences at stake for them, were much higher than for other developments. The number of expectations and visions in play – often conflicting – was also higher. We can consider a policy document to be, as with a standards document, both an envisioning and an institutionalising device. The fact that two policy documents for energy in Kenya – the MOE and the Policy Dialogue versions – were competing served to intensify the political struggles. Of course, the MOE felt that their vision had more legitimacy, being an agent of an elected government, but the Policy Dialogue could also claim legitimacy, as it had involved a much wider range of stakeholders than the MOE process. The outcome, as expressed in the Energy Act 2006, was a compromise between these competing visions, whereby PV retained some recognition, as we have said, through the exercise of the power of parliamentarians.

Of course, the parliamentarians did not act spontaneously. Niche actors deployed socio-technical expectations in order to recruit their support, and the parliamentarians, having experienced this second-order learning, began to adopt the detailed vision, expressed in the Policy Dialogue document, with the help of actors such as ESDA and GTZ. And ESDA and GTZ themselves experienced second-order learning as a result of their 'success' in influencing the Energy Act. For Otieno at GTZ, and the parliamentarians concerned, that learning was expressed in the formation of the Parliamentary Network on Renewable Energy and Climate Change, that is, the forming of a partial vision that policy outcomes on renewable energies could be influenced through parliamentary actors, further envisioned by employing ESDA to conduct seminars for those actors.

We can see that the interactions of niche actors with the regulatory and policy regimes were important for niche development in a number of ways. There were the kinds of outputs we might expect: technical standards from interactions with the regulatory regime and an Energy Act reflecting some of the interests of the niche from interactions with the policy regime. But there were other outcomes that were significant for niche development. The work on the standards committee stimulated the formation of KEREA. This has the potential to further articulate the networks within the niche and connect to networks beyond, as well as being an industry voice for interactions with government. It also created a code of conduct in addition to the technical standards, which could be important for institutionalising practice among the niche actors. And the policy experience was rich in learning for some of the key actors in the PV niche, particularly in terms of how to lobby and influence the policy regime.

However, the more recent developments in energy policy-making in Kenya have reversed the fortunes of the PV niche to some extent. Actors in the PV niche do not necessarily share the now-dominant expectation of low carbon development using – in particular – geothermal energy. Instead, there is a powerful network of actors who do share, or could easily share, this expectation and they are driving the policy-making process, whether because of institutional legitimacy (MOE), control of resources (donors, large industry) or the power to effect political change (grid-connected consumers). Those PV niche actors with expectations and visions centred on SHSs and SPLs are unable to wield any countervailing power against this dominant array of forces. In a sense, they are being forced into accepting an expectation of the private sector-led PV market, one that does not incorporate the need for protection and nurturing. Instead, it is one that must accept discipline and taxes. There are probably still enough donors who hold expectations of SHSs and SPLs that further nurturing of the niche is likely, but the discipline imposed by the expectation embedded in the new regulations is stark for some actors. It begs the question as to whether the niche networks will now fragment as the poorer actors find they are unable to pay for training and licences, or whether the market is vibrant enough for them to continue making a living from PV installations. The new markets – discussed below – might offer these livelihoods.

New markets: pico-solar and module assembly

We have already mentioned Lighting Africa several times, and discussed their attempts to introduce minimum standards for SPLs. But the programme did much more than try to address the issue of quality in the pico-solar market. Here, we discuss the other interventions Lighting Africa implemented in Kenya, and attempt to demonstrate that these together could be considered a systemic approach to market creation and development. Alongside these interventions, the market for pico-solar products in Kenya has grown rapidly and has seen the entry of a large number of new private sector actors and products. While it would be problematic to attribute this market growth entirely to the Lighting Africa interventions, there is certainly a strong correlation. Nevertheless, as we will see in the discussion below, without Lighting Africa's advocacy, it could be argued that the variety of new pico-solar products now available would not have been developed, and so it is unlikely that such a market would have emerged. Indeed, the evidence discussed below suggests that in-depth research focussed on the Lighting Africa programme – or similar interventions – could yield important insights for pro-poor low carbon development in general.

This section also briefly discusses the establishment of Kenya's first PV module assembly plant; an interesting development, in that it suggests the Kenyan PV sector is moving on a trajectory that could see it capture more of the PV value chain. It is still too early to assess the extent to which this is likely, but several PV actors consider the plant to be successful to date, and it appears to be employing a careful strategy to build confidence in its products in the East African region.

Targeting the bottom of the pyramid: the pico-solar market

In September 2007, the IFC launched the Lighting Africa programme. This was a collaboration between the IFC and the World Bank, with a range of donors in support, intended to build on previous market development interventions such as the Lighting the Bottom of the Pyramid (IFC 2007a), a GEF-supported programme (Lighting Africa 2009, p. 2; Anonymous 2013b). The first phase involved a global call for project proposals aimed at developing new lighting products and delivery models for Africa's large unelectrified rural off-grid lighting market (Development Marketplace 2007). The call was launched in partnership with the World Bank's Development Marketplace initiative, which had already been in operation since 1998. In Kenya, as we have discussed in Chapter 4, there had already been several (unsuccessful) attempts to bring PV-powered lighting technologies to the poor, and recent attempts to develop pico-solar products (e.g. the radio and phone charger marketed by Fred Migai). The hope with Lighting Africa was that recent advances in performance of key technologies – especially LEDs – could be harnessed to provide cheaper and better lighting for the bottom of the income pyramid (BOP).

Grants of up to USD 200,000 were available for each successful proposal, and 16 were selected from the more than 400 proposals received, four of them to be implemented in Kenya (Lighting Africa 2008c, p. 7). Three of these involved PV (winning company in brackets): consumer finance scheme for SHSs (ESDA); transfer of LED lantern assembly from India to Kenya (Thrive) and a rent-a-light scheme (Solar World). The grants were awarded at a ceremony during the first

Lighting Africa conference, held in Accra from 6–8 May 2008 (Lighting Africa 2008c, p. 6). Since then, Lighting Africa conferences have been held in Nairobi (2010) and Dakar (2012), during which awards were given for a selection of 'outstanding' lighting products already on the market rather than from a competition such as the Development Marketplace (Lighting Africa 2011; Lighting Africa 2013a). And, in October 2015, a fourth conference took place in Dubai, although it is not clear that any product awards were given this time (GOGLA and Lighting Global 2015).

By the time of Lighting Africa's second-year progress report in 2009, the programme had begun activities on many fronts, including market research in several countries, product testing and the development of quality assurance methodologies, identification of financing needs throughout the value chain, knowledge-sharing and self-evaluation and moves to identify policy constraints by researching the policy environments in several countries (Lighting Africa 2009). For Kenya, by the end of 2008, there were already highly detailed qualitative and quantitative market assessments (Lighting Africa 2008a; Lighting Africa 2008b). And much more research followed, including on products available in Kenya, product testing, and a review of the policy environment and policy actors (see the Lighting Africa website[9] for these reports).

In 2009, Lighting Africa began its market development interventions in Kenya (the other pilot country being Ghana). Above, we discussed its quality assurance activities, in which the programme developed minimum performance standards for SPLs and a testing methodology. This also fed into one of its other activities, which was to influence policy at the national level. Also at the national policy level, and through the World Bank relationship with Kenyan policy makers, it hoped to address the issue of import duties and taxes on PV products. As we have seen, this has been an uneven and unpredictable experience, with taxes being levied and then removed and then levied once more. For many of the private sector interviewees, the issue of taxes and duties has been particularly vexed. According to their testimony, the reintroduction of 16 per cent VAT in October 2013, for example, has severely reduced sales of their products – in some cases as much as 30 per cent. Others point to 'perverse' incentives created by import duties, where complete PV systems are duty-free while components are not. This, they claim, forces local companies to import systems rather than purchasing components for assembly in Kenya. That is, the duties are not helping to develop local value chains. In the opinion of some interviewees, the role of actors such as Lighting Africa should be to lobby for the removal of such policies rather than to create a market.

Nevertheless, many of our private sector interviewees noted that Lighting Africa had been helpful in its market development activities, and some actors in the market were beneficiaries of the initial grants to help get products and delivery models started. This brings us to the other three aspects of the Lighting Africa programme: (1) business support and access to finance; (2) access to finance across the supply chain; and (3) consumer education (Anonymous 2013b).

In short, business support includes identifying potential dealers in rural areas and connecting them with suppliers and organising trade fairs to bring suppliers and buyers together (the conferences mentioned above). Access to finance for business includes enabling credit so that companies can increase their stock of products. Lighting Africa did not itself provide finance and it is unclear whether the programme was able to deal successfully with this aspect of the intervention.

However, when considering access to finance across the entire supply chain, there seems to have been more success. In some ways, this intervention was similar to PVMTI in that it was concerned with finance on both the supply and demand sides of the market. But it differed from PVMTI in important ways. First, Lighting Africa was not lending any money. Instead, it helped to develop two models of finance. One – the bulk-buyer or corporate outreach model – began with Unilever, which had over 10,000 employees in the tea sector who could be customers. Unilever experimented with a 'check-off' system of payments, whereby each employee who was buying a light had a certain amount of money deducted from their salary each month. In essence, this was a hire-purchase model, except that Unilever was acting as the loan-agent rather than a third party. The other model was for Unilever to lend money to its associated Savings and Credit Cooperative (SACCO), which could then lend to its members. Again, the basic model is familiar in Kenya, except that Unilever could charge a lower interest rate than a bank would do, and so the SACCO could pass the saving onto the customers. The second important difference was that PVMTI was constrained to lending a minimum of USD 500,000 to a micro-finance institution (MFI), whereas the Lighting Africa experiments could lend much smaller amounts – more appropriate in the context of rural Kenya. It is claimed that both these models have been successful and have been adopted by other large companies and SACCOs (Anonymous 2013b).

The last aspect of the Lighting Africa intervention was consumer education. This was considered the most challenging of the interventions and expensive to implement. It required understanding of what the consumer does and does not know, and involved running forums, road shows, meetings, and more. This was also an aspect in which Lighting Africa learned by doing, evolving its approach with experience. For example, it became clear that just raising awareness could mean disappointment for potential customers. Once they were interested in the idea of SPLs, many wanted to purchase them immediately. If there were no dealer to sell the products, then the customer would likely be dissatisfied. To avoid this – it is claimed – Lighting Africa combined awareness-raising activities with its retail outreach in the current target area. It also often had an MFI with it. For those who could not afford to buy immediately, Lighting Africa developed a text message service whereby the customer could send a blank message to the number some time later and receive a list of approved products. This was available in the relevant local language. While this appears to have been a successful service, it is not clear whether it will be continued now that Lighting Africa has finished its interventions in Kenya.

The programme innovated its marketing campaign in a number of other ways too, although there is no space to detail everything it did. The point to make here is that it did so in response to its greater and evolving understanding of each context into which it moved. In 2012, it was awarded a prize by the Marketing Society of Kenya for the 'best experiential campaign in the NGO/Government category' (Lighting Africa 2013b, p. 1). According to the same report, by the end of 2012, Lighting Africa had run over 1100 forums and 190 road shows in Kenya, reaching an estimated 260,000 people.

Up to the official completion of its Kenya pilot phase in July 2013, the programme continued to engage in the combination of interventions described: aggressive and roaming awareness-raising campaigns, quality assurance of products, setting-up of a product quality testing facility, training of technicians, capacity-building for business development and for finance institutions, lobbying of policy-makers on regulations and building networks of actors to encourage the flow of information. While it is difficult to determine the extent to which outcomes can be attributed directly to these efforts, the programme does make a series of claims. One of these is that, at the time of the field research, the annual Kenyan market for good-quality pico-solar – alone – had grown to sales of over 100,000 products (Anonymous 2013b). This may have been an underestimate, given that a recent updated survey in three towns in Kenya supports the notion that the market for small off-grid lighting products has expanded extremely rapidly since 2009. Citing figures from the Kenyan Ministry of Energy and Petroleum, Turman-Bryant et al. (2015, p. 6) report that there were accumulated sales of 2.3 million quality-assured pico-solar products as of June 2015.

Not all private sector actors have entered the pico-solar market because of Lighting Africa, as can be seen from the discussion in this section. However, it is unlikely that these interventions have been completely ineffectual, and it is clear that many actors would not be aware of the products or where to buy them (both dealers and customers) if Lighting Africa had not intervened. Still, it is remarkable that, as with the rest of the long history of PV described in this case study and the persistent involvement of many donors over this time period, the PV market in Kenya continues to be described by most observers as 'unsubsidised'. The Lighting Africa programme in Kenya alone cost USD 5–6 million (Anonymous 2013b). The whole pilot programme, inclusive of other countries, is in excess of USD 12 million.

Moving on up: Kenya's first solar module assembly plant

There had been at least one attempt in the past to establish manufacturing of PV modules in Kenya, although this attempt fell apart following the post-election violence in 2008 (Disenyana 2009). The intention had been for a Chinese company to start a joint venture in Kenya to manufacture amorphous modules in Nairobi. It is possible that other attempts have been made, but none appear to be documented. However, Ubbink EA began assembling polycrystalline modules in Naivasha in August 2011 (Oirere 2012), the result of a long process that could bring more value-added to the Kenyan PV niche (Kimuya 2013, interview).

The process was apparently initiated around 1999 or 2000 by Chloride Exide (Mabonga 2013, interview). They had already been sourcing modules from a

Dutch group of companies (Ubbink BV) for many years. According to Kimuya (2013, interview), the two companies first tried to establish the plant in Ethiopia, but the policy environment was not conducive. Eventually, they decided to open in Kenya. This process took almost a decade, and it is not clear what the explanation for this is. Kimuya suggests it may have been a combination of political and bureaucratic difficulties, as well as the task of identifying suitable personnel. In any case, Ubbink East Africa – a joint venture between Largo Investments (who own Chloride) and Ubbink BV (Centrotec Sustainable AG) – was officially registered in Kenya in 2009–2010 (Mabonga 2013, interview). Three technicians were then sent to the Netherlands for one month of training, and they trained six more upon their return to Kenya. This continued and, at the time of the research, there were 78 Kenyans trained to operate the machines in the assembly plant (Kimuya 2013, interview). Naivasha was chosen as the location because of the lower cost of land and, being on the northern corridor, it offers good transport links for the main markets. The largest market is Western Kenya, and Kenya is the largest country market that Ubbink EA serves.

Half of the investment for the plant (said to be USD 3 million: Oirere 2012) was provided by the Dutch Government and the other half was shared between Chloride and Ubbink BV (Stuart 2011). The factory first produced 180 kW of modules per month, but this has risen to about 250–300 kW as a result of continuous improvements to the production process (Kimuya 2013, interview). They produce a wide range of sizes – from 13 Wp up to 240 Wp. The most popular module size is 40 Wp, which is considerably larger than the most popular module size in the market in the past. This used to be 12 Wp (van der Plas and Hankins 1998), but Kimuya suggests that the falling price of PV has meant that people are able to buy larger modules and so meet more of their demand. The preference in Tanzania is for 50 Wp modules, and it is 80 Wp in Uganda, where the subsidy for projects is generous.

Kimuya (2013, interview) claims that Ubbink EA has built a solid reputation in East Africa by inviting distributors, dealers, retailers and technicians to visit the factory where they also receive basic PV training. During these visits, they are shown around the production process and they talk to the staff. This way, it is claimed, they build trust in the company and the product. There does appear to be a general assessment among PV niche actors that the company is succeeding (Newell *et al.* 2014), and they are well connected into the PV actor networks in the region, not least through the Chloride Exide contact.

With Ubbink established as a strong name in the region, they were, at the time of the field research, considering diversifying the production to include goods with PV embedded, such as solar radios, lanterns, TVs and street lamps. This would make sense in the medium term, as any attempt to move into the manufacture of cells would require much higher investment risk and lengthy capability-building efforts. Still, as we discussed in Chapter 2, the Innovation Studies literature provides evidence from many examples around the world demonstrating that assembly is the first of many steps in the direction of building more complex manufacturing and innovation capabilities. Recently, it appears, Ubbink have begun to assemble products with embedded solar, and this is being done under licence from a German company,[10] *fosera*. At the time of writing, they had at least one solar lighting kit listed on the KEBS website, indicating that it has the KEBS Standardization Mark, and there were four quality-assured products shown on the Lighting Africa website.

Socio-technical analysis of new markets

Lighting Africa and the pico-solar market

In some ways, the Lighting Africa programme could have been designed on the basis of niche theory. It has made huge attempts to recruit actors to its network by collectivising an expectation: it has evolved through learning, attempted to institutionalise many socio-technical practices, encouraged a diversity of experiments in different contexts, nurtured and protected and acted as an innovation system builder. Ironically, the vision guiding this behaviour is one of entrepreneurs seeking profit in a free market and thereby providing clean lighting services to the poor. Of course, this same vision – or expectation – has been promoted throughout the history of the SHS market in Kenya, and it has proved to be a successful vision for recruiting public resources to assist this private market. This is not to argue that these resources have been poorly used. Our argument is quite the opposite, and the Lighting Africa programme goes some way to demonstrating that the systemic approach it has taken is potentially both faster at delivering energy services to the poor and more sustainable than a light-touch free market approach.

From the early documentation that led up to the implementation of Lighting Africa, it is clear that there was a concerted effort to build a strong network of actors who could adopt an expectation of pico-solar lights for the poor in Africa (IFC 2007a, p. 6). This effort began more than two years before the launch of the Development Marketplace competition and the IFC consulted over 190 actors in the process. This network was further enhanced through the launch competition and new lighting product and service ideas generated. The winning 16 ideas were then given protection with the grant money. Because they were implemented in a wide range of contexts, this could generate valuable learning about these new socio-technical practices in real-world settings.

Further learning was enabled through the market surveys – both quantitative and qualitative – and policy environment studies commissioned by the programme across several African countries. These were all shared on the Lighting Africa website. As we have seen in earlier sections of this case study, this detailed articulation of contexts is essential for the development of socio-technical visions and their further collectivisation. As actors adopt shared expectations and visions, so they focus on solving similar problems when trying to realise those visions. This raises the chances that those problems will be solved or that new expectations will be stimulated. A simple example in the Lighting Africa experience in Kenya was the discovery that many importers did not actually know what constituted a good quality solar lantern. This led to the testing of a range of lights with users and the eventual identification of a minimum acceptable standard. With this standard codified, Lighting Africa was able to go back to the manufacturers and tell them in what ways their lights needed to be improved. The second round of tests then showed that the manufacturers had actually responded to this.

Of course, this codifying of a minimum acceptable quality then led to the further development of related standards and a testing laboratory. Subsequently, Lighting Africa initiated the process of trying to institutionalise these standards globally, forming other actor-networks in the process. And these practices are now in the process of being institutionalised in Kenya: the standards have been officially adopted and the national testing facility is able to conduct some degree of SPL screening. Not all actors have adopted their expectation of quality but Lighting Africa has started a process from which to generate learning that could lead to others eventually adopting some version of it. In any case, this expectation of quality has stimulated a diversity of innovations in SPLs, more than 40 of which have been accredited.

There have been innovations in other aspects of the niche too. Working with others, Lighting Africa has experimented with micro-finance models and with marketing techniques. Others have entered the niche with their own experimental business models incorporating, as discussed in Rolffs et al. (2015) on consumer finance, ICTs and PV systems. Nurturing has been given to some of these actors through provision of marketing and bearing the risks and expense of finding demand and connecting it with supply. In a country like Kenya, where many people live in remote rural areas, such activity is time-consuming and expensive. While not everything has worked for the programme, and not all actors are satisfied, it has done the work of building elements of an innovation system around pico-solar lighting products and business models.

Ubbink module assembly plant

It is too early to say much about the Ubbink EA assembly plant. Clearly, the existing relationship between Chloride Exide and the Dutch module supplier was important in initiating the idea to establish such a plant in East Africa. And, considering the decade this took to realise, this relationship must have been quite strong. We might expect that there were significant amounts of learning during this establishment process, but it is impossible to say at this time what the content of such learning was.

We can see that the institutional environment seems to have played a role in attracting Ubbink to Kenya rather than Ethiopia, although we cannot be clear about the details of this. However, there is suggestive evidence that there may have been close communication – perhaps even lobbying – between the Kenyan Government and the joint venture investors. Mabonga (2013, interview) hinted that Chloride Exide influenced the government in terms of policy, and the budget of 2011 – announced in June, ahead of the assembly plant opening in August – included the removal of duties on the raw materials for making solar modules (KPMG 2011, p. 7). Once again, here is evidence of the complicated relationship between the policy regime and the PV niche.

There is also further evidence of subsidy in the PV niche. This time it was in the form of the Dutch Government providing 50 per cent of the investment for the assembly plant. We can speculate that making such an investment in Kenya would have been seen as risky by those in the joint venture, and so this subsidy can be understood as some level of protection against this risk. It would also, of course, be of potential interest to the Dutch Government itself, in terms of representing the interests of Dutch industry, depending on whether the market for modules becomes large.

But there have been other benefits for the Kenyan PV niche. At least 78 Kenyans have been trained in the production process to assemble polycrystalline PV modules. And they have developed their capabilities in-house to improve this production process. This is an instance of what Bell (1990) refers to as the development of production capabilities. The local supply chain has begun to capture more of the value-added available from the PV market, and parts of the regional supply chain have become more interconnected. Furthermore, if they had not been trained before, those who have visited the plant (at Ubbink's expense: Kimuya 2013, interview) have gained at least some basic PV skills. And they have had the opportunity to meet others in the regional PV networks. Finally, there is the prospect of new locally-sourced pico-solar (and other) products being developed, especially if they can exploit learning opportunities from the experience of assembling the *fosera* pico-solar products.

Conclusion

In Chapter 4 we charted the arrival of PV into Kenya in the late 1970s and early 1980s. The technology was used for community and commercial services at that time, but it also made equipment available that was then used by others in the country. Notable, from the perspective of the theoretical ideas developed in this book, was that Harold Burris began to use the technology and experiment with business ideas. Then, together with Mark Hankins, he exploited the availability of PV for the installation of systems in several school projects that spawned the idea of solar home systems. These SHSs were then taken up by other private sector actors, beginning with those who were already supplying PV equipment in Kenya. Soon, the SHS market began to flourish and Mark Hankins started to seek donor-funding to experiment with many ideas for product development and business models, including many actors in the market in these projects. The outcome was a strengthening niche and growing market.

Powerful international development actors then started to become interested in this phenomenon and resources began to flow more readily, assisting Hankins and many others to develop the niche further. As the niche developed and the market grew, so more actors entered and gradually specialised in particular roles. Now, with the advent of technical and economic improvements in LED technologies, a new market for pico-solar products has developed, fostered by Lighting Africa. Niche actors have begun to interact with policy regime actors, and have scored some successes in terms of influencing policy to encourage further market growth. However, these relationships have been unstable and they appear to be in decline at present, as the policy regime turns its attention to exploiting the vast geothermal energy opportunities in Kenya.

Nevertheless, the activities of actors such as Hankins, his company EAA, and others such as KEREA and Lighting Africa, have helped the Kenyan PV niche to accumulate many elements of a nascent socio-technical innovation system. And the recent establishment of a solar module assembly plant is suggestive that the Kenyan PV niche is opening a trajectory of development that could lead to much more complex capabilities that might result in the emergence of more sophisticated local innovations. A number of key conclusions of relevance to policy, practice and academic theory can be drawn from these insights.

Rebutting the private sector-led development narrative

The analysis above clearly demonstrates how the success of the Kenyan market for off-grid solar electrical services can be attributed to a range of targeted interventions by key actors over time. These contributed by building technological capabilities where gaps existed and putting in place vital parts of a functioning socio-technical innovation system around off-grid solar in Kenya. This perspective firmly rebuts the received wisdom of many people who often comment on the case of solar in Kenya, showing that it most certainly *was not* a simple case of free market forces driving success. Many of the key actors involved were private sector actors. However, they acted both with and without public funding and support, pursuing a range of capability-building activities that served to put in place the components of a functioning socio-technical innovation system that previously did not exist. These activities were separate from and in addition to conventional rent-seeking activities and were integral to providing the basis for long-term, sustained development of the market for off-grid solar in Kenya. Below we draw out some of the key points from the analysis in Chapter 4 and Chapter 5, demonstrating how they support a socio-technical innovation systems perspective (as articulated in Chapter 2 and Chapter 3), as opposed to the Hardware Financing and Private Sector Entrepreneurship framings that currently dominate research and policy in this field (as articulated in Chapter 1). We conclude the chapter by articulating key policy lessons that can be drawn from the analysis.

Great expectations

Explaining the evolution of the Kenyan SHS and, more recently, SPL markets is perhaps best begun by examining the use of socio-technical expectations. From the emergence of the SHS market in the mid-1980s up to the mid-2000s, the dominant expectation was of a market for PV systems. This increasingly shared expectation guided the search and problem-solving activities of a range of actors – public and private, international and local – on the specific issues relevant to market development. By focussing problem-solving on issues of relevance to all SHS market actors, any lessons generated were widely and readily applicable. Moreover, as many of these problem-solving activities were funded by donors, the lessons tended to be made public through reporting and through discussion in various forums (such as workshops and other meetings, and likely by word-of-mouth through the well-integrated actor networks in the Kenyan PV niche). Private sector actors were then able to make use of this learning in further focussed activities of their own that helped to grow the market.

From about the mid-2000s, a variant of this PV market expectation – based on solar lanterns – began to take hold. Solar lanterns had been available for many years, and had been the subject of various experiments in Kenya, but there were several technical and economic characteristics that made adoption by poorer users difficult; the lanterns offered limited functionality and reliability at prices only slightly below those of the smaller SHSs. But technical improvements in lighting technology – especially LEDs – meant that there was an opportunity to later revisit lanterns as a solution to lighting services for the poor. The IFC began work on constructing an expectation that married these technical possibilities with the rhetoric of the bottom of the pyramid, and it actively recruited actors globally to this expectation. Bolstered by the large network of actors so recruited, the Lighting Africa programme began operations in 2007, having persuaded the GEF and others to provide substantial funding. At the time, however, there were few (if any) lighting products designed specifically for poor African users, and so Lighting Africa stimulated the design of a range of products through its international competition for grants.

The grant competition recruited more actors to this new expectation and provided protection – in the form of the grants awarded – for a number of experiments with different products and delivery models in various contexts. Several other niche-development activities – similar to those seen in regard to developments around SHSs – were then implemented when Lighting Africa began direct interventions in Kenya in 2009. These included articulation activities (descriptive of market demand and problems in the market, connective of the actors in the supply chain and with demand), building actor networks, socio-technical learning and sharing lessons, and institutionalising practices from the use of pico-solar products through to formalising performance standards and testing. In the meantime, as this expectation was widely deployed and adopted, and as it became increasingly refined in a particular socio-technical vision, other private sector actors have been attracted to the Kenyan pico-solar market, where they have experimented with a wider variety of products and business models. Whether this widening variety of actors and business models can be causally attributed to the activities of Lighting Africa is difficult to establish, but the rise of the pico-solar market has a striking correlation with these activities. Certainly, it is an area worthy of further research in order to learn more comprehensively from Lighting Africa's experiences. This includes learning from the relative success in other countries, including the parallel pilot project that Lighting Africa implemented in Ghana. Some commentators report the initiative meeting with more limited success in Ghana (Urama 2014). Important policy-relevant insights could be gained from research comparing experiences in the two countries, especially if analysed through a socio-technical innovation systems lens. This could test the hypothesis that the uneven success of Lighting Africa across Ghana and Kenya is the result of a better-developed socio-technical innovation system in the latter, as charted in the detailed historical account given in Chapter 4 and Chapter 5.

The political nature of Socio-Technical Innovation System Building

Another important feature of the use of expectations – visible in the evolution of the PV niche, and both the SHS and SPL markets – is their political nature. That is, when an actor deploys an expectation, they are attempting to persuade others to adopt it. As per pathways thinking (Leach *et al.* 2010a), this involves the construction and use of a narrative that defines the problem, and identifies an intervention that will solve that problem. In this case, of course, the problem has been the long-standing issue of electricity access, especially for poorer groups in rural areas. For much of the evolution of the Kenyan PV niche, the intervention suggested in the narrative was to address market failures so that private sector actors could more easily sell good quality SHSs to customers in a promising free market. Hankins (and subsequently EAA) was particularly important in constructing and deploying this narrative, persuading many to adopt the expectation of a PV market in Kenya by emphasising that rural people would benefit from better development outcomes. It is fair to say that hundreds of thousands of rural people have indeed benefited from electricity access via solar, and many others have benefited from the profit and employment associated with the SHS market.

But it became increasingly clear that poorer groups in rural areas were not getting access to electricity from solar and that it was difficult for them to do so. Attempts to solve this problem within the SHS market expectation – by using micro-finance – largely failed (most conspicuously in PVMTI), and it could be argued that this stimulated the second-order learning that created the pico-solar expectation. However, the early experiments with pico-solar products were unsuccessful, especially as the first solar lanterns were too expensive and provided limited and unreliable functionality. It was not until techno-economic advances in LEDs became available that solar lanterns (and subsequent varieties in functionality) could be seen as viable, and a new narrative could be constructed that convincingly included private sector provision of electrical services for the poor. While there were some actors in the Kenyan niche experimenting with pico-solar products – such as Leo Blyth and Fred Migai – they were not politically active. That is, they were not deploying a pico-solar expectation or narrative and so had little support among the established SHS actors, who did not take these pico-solar products seriously. It took a powerful actor – the IFC – to create widespread interest[11] in, and adoption of, a pico-solar expectation.

Here we see the operation of power in how narratives influence the direction of development. Actors such as EAA (Mark Hankins' company) were successful in attracting resources to enable development of the SHS niche and growth of the market. However, in the main, these resources were made available in small quantities that enabled only small-scale interventions and experiments. The only exception to this was PVMTI, an intervention that failed (in its primary goal) because it was based on a misunderstanding of the problems in the SHS niche, pursuing, as it did, interventions that can be characterised along the lines of the Hardware Financing framing introduced in Chapter 1. As a consequence, perhaps (and in contrast to the SPL market in Kenya) the SHS niche has taken decades to develop and still faces many problems; chief among them, according to actors in the niche, it would seem, is the issue of poor quality. In contrast, the IFC is a powerful actor and has had a relatively large amount of money to bring to the development of the pico-solar market. Moreover, as a globally credible actor, they were able to persuade many others to adopt a pico-solar expectation even before the Lighting Africa programme had been awarded funding.

Neither the SHS nor pico-solar actors have, however, successfully persuaded those powerful in the Kenyan policy regime that they should adopt any PV expectations. There are some warm words in policy documents, but many PV actors claim that, in reality, there is significant resistance to solar among a few powerful figures in the energy sector. Some actors attribute this resistance to close relations between these powerful figures and the fossil fuel interests in Kenya. Others point to a preference for large projects rather than the small and highly distributed projects that are inevitable with SHSs and SPLs. The reasons for the preference for large projects could range from more lucrative corruption opportunities to the need to realise huge increases in electricity capacity to drive economic growth. Whatever the nature of this alleged resistance, the solar niche does appear to be suffering material effects. The most recent manifestation of this is the imposition of taxes on PV products and the damaging impact this is having on sales, according to those in the pico-solar market in particular. Although we cannot be sure of the specific reasons for what appear to be moves against PV in Kenya, it is clear that they cannot be explained purely in political terms. Rather, if we are to understand them, then we also need to attend to the political economy of such developments, as demonstrated by Newell et al. (2014).

The relationship between niche and regime actors has been somewhat ambivalent in practice through much of the history of the PV market in Kenya. This is not to suggest that all policy actors are negative towards solar; there are many who have adopted the expectations that are in play. And there have been policy innovations that might help to support further nurturing of the niche. For example, the development and adoption of a national PV curriculum could have important capability-building benefits for the long-term health of the niche. The introduction of PV regulations could also address the long-running issues of poor quality, although there could also be detrimental effects too. For example, those technicians who cannot afford the required training and licences will be forced to either cease work in the PV niche or risk becoming criminalised. Capacity in the PV niche could then decrease and the niche networks fragment, potentially weakening niche development rather than strengthening it. But, beyond niche-regime interactions, there are also continuing attempts to both understand and improve user practices following the adoption of either SHSs or SPLs. In sum, there has been a multiplicity of institutional developments throughout the niche-building process, ranging from informal user practices to the highly codified PV regulations. This work is incomplete, and it is likely to be necessary for many years to come as new products and business models are introduced.

Socio-Technical Innovation System Building

The final conclusion we can draw from the analysis in this research relates to our hypothesis that key actors have undertaken Socio-Technical Innovation System Building activities in the Kenyan solar niche, and that these activities explain the relative success of the SHS and SPL markets. This involves work that contributes to increasing technological capabilities, connecting actors across the innovation system and doing this in ways that both account for the social practices of potential technology users (poor women and men) and are able to compete with existing socio-technical regimes in energy production and consumption.

We have referred to the technological capabilities that underpin innovation systems as skills, knowledge and linkages between actors throughout an economy. Our socio-technical focus also places emphasis upon linkages across society, given that we are interested in the provision of electrical services in that society. Here, we can see that actors such as EAA not only deployed expectations that recruited resources and other actors, and guided problem-solving activities; they also drew lessons from projects that informed subsequent interventions, helped devise and deliver training to improve skills, shared knowledge about the market, linked actors together to develop supply chains raised awareness of PV among customers and linked them to these supply chains and lobbied the policy regime for a more conducive institutional environment. Taken together, these lines of activity have helped the niche accumulate elements of a socio-technical innovation system around PV. In the case of the SHS market, EAA (and its subsequent incarnations) were clearly the most important of these innovation system builders for many years. As the niche strengthened, this role has become more distributed (others such as SolarNet, KEREA, KEBS, etc., have taken on more specialised but mostly complementary roles). In the case of pico-solar, Lighting Africa has been the most important actor, although it has benefited from the existing niche networks and those with relevant skills, knowledge and linkages.

Without such capabilities in place, it should be clear that the Kenyan 'free market' in solar would not function. Moreover, the need for capability-building has not diminished. Three in every four Kenyans still have no access to electricity, whether via solar PV or otherwise. Apart from the capacity constraints that actors say the market currently faces, new products and business models continue to be the subject of experimentation, and these may well require new specific capabilities. Even without new products and business models, international climate policy instruments, such the CDM, have not yet been exploited in the Kenyan solar niche. If they were to be – along the lines that China, for instance, has exploited the CDM (Stua 2013; Watson *et al.* 2015) – then there would need to be other specific capability-building efforts if these new products and business models are to be successful. Given the continuing ambivalence of the Kenyan policy regime to small-scale PV-powered electrical services (Newell *et al.* 2014), there remains considerable political work to do to persuade regime insiders to adopt PV expectations.

Alongside these political needs, there is still a tremendous amount of niche-development work to do in relation to both SHSs and pico-solar. The history of the PV niche in Kenya suggests that this kind of work needs to be done by coordinating actors – by actors who are positioned to structure practices and to build socio-technical innovation systems. The recently operational Climate Innovation Centre could be such an actor in Kenya. Where niches already exist to some extent, there might be a need for specialised and complementary roles so that the CIC might better focus on particular aspects of the innovation system. Where there is little or no niche activity, the CIC might need to take on more of the activities necessary to develop the niche, much as Lighting Africa has done in regard to pico-solar. Whatever role the CIC takes, it is clear from the evidence and analysis here that interventions need to be systemic rather than narrow or piecemeal, be patient and involve a wide diversity of interested actors.

Having articulated this detailed innovation history charting the emergence of the Kenyan off-grid solar PV market, in the final two chapters we attempt two things. First, in Chapter 6, we explore the policy implications of the insights yielded in Chapter 4 and Chapter 5. In doing so, we compare examples of the dominant Hardware Financing and Private Sector Entrepreneurship framings introduced in Chapter 1 with a policy approach (Lighting Africa) that looks much closer to the alternative Socio-Technical Innovation System Building framing introduced in Chapter 2 and Chapter 3. Through this analysis, we emphasise the relevance of a Socio-Technical Innovation System Building approach to and practice around sustainable energy access and low carbon development more broadly. This includes suggestions for how international climate policy might be revised in ways that could facilitate such Socio-Technical Innovation System Building.

Second, in the concluding chapter (Chapter 7), we summarise the overarching theoretical, policy and practice insights yielded via the analysis in this book. Importantly, we also address the critical question of whether these insights only apply to the case of solar PV in Kenya. This introduces insights from work in China, India and Tanzania, and plays a critical role in establishing the extent to which the ideas in this book might both be transferable across different national and socio-technical contexts and provide a useful conceptual basis for future research and interventions through policy and practice.

Notes

1 There were interactions of some kind before this, but they were mainly at seminars and workshops, such as the 1992 Regional Training and Awareness Workshop (Kimani 1992).

2 EAA became connected with ESD, a company in the UK, starting around 1998 and changed its name to ESD sometime in the early 2000s. This then became ESDA some time later and, around 2007, Camco Advisory Services.

3 Otieno was invited by the committee to observe (Otieno 2007, interview).

4 It is not just fear, of course. The parliamentarians have institutional power to accept or reject policy (Otieno 2007, interview).

5 The components listed in the regulations are (with warranty periods): charge controllers and regulators (10 years), inverters (10 years), batteries (1 year), light bulbs and LEDs (1 year), panels (20 years) and light fittings and devices (2 years) (ERC 2012, p. 18).

6 The Ministry of Energy has recently been renamed the Ministry of Energy and Petroleum.

7 For Burris, of course, this was an issue from the outset. And Hankins was an early recruit to Burris' vision.

8 Of course, KEREA covers other renewable energy technologies as well as PV, and the code of conduct is for all its members.

9 A wide range of materials is available on the Lighting Africa website: www.lightingafrica.org/

10 According to the *fosera* website, the company has its main factory in Thailand but has other assembly lines around the world. The Ubbink assembly line began operation in 2014 and is producing 70,000 units annually. See www.fosera.com/company/assembly-network/kenya-naivasha.html

11 It is not clear from this research who, specifically, initiated the idea of what became the Lighting Africa programme and, indeed, why they adopted such an expectation in the first place. As noted above, further research into the emergence of this narrative and mobilisation of resources around it could provide valuable lessons for policy and theory.

6

LEARNING FROM THE KENYAN SOLAR PV INNOVATION HISTORY

The detailed historical analysis in Chapter 4 and Chapter 5 demonstrates the ways in which the development of the off-grid solar PV market in Kenya resembles processes of Socio-Technical Innovation System Building. In this penultimate chapter we focus specifically on the implications of these observations for policy and practice. We begin by revisiting the dominant policy framings introduced in Chapter 1, namely Hardware Financing and Private Sector Entrepreneurship. We then discuss how approaches based on these framings contrast with a Socio-Technical Innovation System Building approach and how the latter holds more promise in terms of supporting transformative policy interventions with greater potential to address the sustainable energy access problematic.

In relation to each framing, we provide an empirical example. We begin by contrasting the Hardware Financing framing of the Photovoltaic Market Transformation Initiative (PVMTI) with Lighting Africa's more Socio-Technical Innovation System Building-oriented approach. We then introduce a new policy intervention that was not covered in the innovation history in Chapter 4 and Chapter 5, as it is too new to be able to chart tangible results. This is the example, mentioned in Chapter 1, of Kenya's new Climate Innovation Centre (CIC) – an infoDev/World Bank initiative with support from the UK's Department for International Development (DFID) and its Danish equivalent, Danida. The CICs represent a good example of a Private Sector Entrepreneurship framing to the problem of sustainable energy access and technology transfer more broadly. While it is too early to provide any detailed empirical evidence, we present what evidence is currently available in the public realm on the Kenyan CIC and discuss the potential implications of the framing it adopts in terms of its likely impact in transforming sustainable energy access, or supporting low carbon development more broadly in Kenya or beyond. This *ex-ante* analysis of the transformative potential of the CIC approach builds on the insights from the historical analysis of the solar PV market given in Chapter 4 and Chapter 5, and the Socio-Technical Innovation System Building perspective these insights support.

Then, having presented quite a critical perspective on the two policy approaches that exemplify the framings we have argued are inadequate (PVMTI and CICs), we end the chapter by putting forward some recommendations for how policy might operationalise a Socio-Technical Innovation System Building approach in order to maximise chances of creating the conditions within which transformations in sustainable energy access are more likely to occur. This draws on lessons from the Lighting Africa example as well as generic insights from our detailed innovation history of solar PV in Kenya. In doing so, we articulate a vision for a new kind of institution with the potential to better support the development, transfer and adoption of sustainable energy technologies in developing countries, especially among lower-income countries where socio-technical innovation systems are less well developed, but where significant opportunities exist to build such systems around sustainable, pro-poor technologies across the board (energy-related or otherwise). We have, elsewhere, introduced this idea in the form of institutions that we have called 'Climate Relevant Innovation-system Builders' (CRIBs) (see Ockwell and Byrne 2015), but we believe it is applicable to other sustainability issues.

Hardware Financing and Private Sector Entrepreneurship versus Socio-Technical Innovation System Building

In Chapter 1, we introduced three alternative framings through which the issue of sustainable energy access – as well as linked concerns such as low carbon technology transfer and low carbon development or green growth – can be understood. Two of these, Hardware Financing and Private Sector Entrepreneurship, tend to dominate current international policy narratives. To recap briefly, these two framings and their associated narratives can be summarised as follows:

- *Hardware Financing*: A Hardware Financing framing of the problem of sustainable energy access is based on a standard Environmental Economics argument. This argument acknowledges, for example, that low carbon energy technologies are not competitive with fossil fuel-based energy technologies because their prices fail to reflect the positive externalities to society of mitigating future greenhouse gas emissions. This leads to the assumption that introducing market-based mechanisms, like the Clean Development Mechanism (CDM), which allow technology hardware suppliers to be paid for these positive externalities (i.e. subsidised), will provide incentives for investment in low carbon energy technologies in developing countries. This argument carries the additional assumption that such investment in low carbon energy hardware will be extensive enough to reorient technological change and economic development in developing countries towards more sustainable directions.
- *Private Sector Entrepreneurship*: A Private Sector Entrepreneurship framing of the sustainable energy access problem focuses at a more micro-economic level. It argues, for example, that low carbon technological change will be driven by

innovation in the private sector and therefore emphasises the need to provide venture capital to entrepreneurial private sector initiatives in developing countries. Again, this argument carries a similar assumption to the Hardware Financing framing, namely that such venture capital provision for sustainable private sector entrepreneurship will reorient countries' development trajectories towards more sustainable directions.

Our central aim in this book has been to demonstrate the limitations of the above two framings for the problem of sustainable energy access, and to provide an alternative. In turn, the alternative should be a framing that is equally relevant to broader issues such as low carbon technology transfer, low carbon development, green growth, etc. The alternative we have been promoting is what we have called Socio-Technical Innovation System Building.

- *Socio-Technical Innovation System Building*: This framing significantly broadens the focus of potential policy interventions, emphasising the need for a systemic understanding of how innovation and technological change can be nurtured through policy. It also emphasises the importance of attending to the social practices with which sustainable energy (and other) technologies intersect, the co-evolutionary nature of innovation and socio-technical change and the existing socio-technical regimes with which sustainable energy technologies must compete.

One potential criticism of this systemic framing of the problem is that the policy prescriptions it generates risk being too context-specific to be practical – that they are not generic enough to be widely-applicable and so are unworkable as policy interventions. Indeed, the attraction of the other two framings, particularly Hardware Financing and market-based mechanisms in general, is that they can be explained with simple narratives and that they come with policy prescriptions that are entirely generic. That is, context-specific issues are irrelevant to the applicability of such policy interventions.

We would dispute these arguments, and so, in this penultimate chapter, our concern is two-fold. First, we provide some comparative examples and more detailed discussion of specific policy interventions that characterise all three of the framings summarised above. This serves both to re-emphasise the limitations of interventions based on Hardware Financing or Private Sector Entrepreneurship and to provide a worked example of Socio-Technical Innovation System Building in practice. Second, we provide a concrete proposal for how Socio-Technical Innovation System Building could be achieved using generic policy approaches, while achieving results that respond to the context specificities that are critical determinants of the impacts of policy interventions across the richly variegated characteristics of different countries (see Chapter 1 for a more detailed discussion of context specificities in relation to sustainable energy access).

Hardware Financing: The Photovoltaic Market Transformation Initiative (PVMTI)

Let us begin with an example from our innovation history of solar PV in Kenya, which fits well with a Hardware Financing approach. In 1998, the International Finance Corporation (IFC) began implementing a project in Kenya that was intended to transform the market by addressing a perceived finance constraint. The Photovoltaic Market Transformation Initiative (PVMTI) made USD 5 million finance available on both the demand and supply sides of the Kenyan PV market, which would be disbursed in loans to consumers and suppliers over the ten-year life of the project (Gunning 2003, p. 81). Finance for customers, it was assumed, would enable them to overcome the high initial cost of PV systems and therefore release pent-up demand. Finance for companies would allow them to purchase in bulk and so reduce their costs, hence lowering prices to consumers. In this way, the initiative clearly adopted the kinds of narratives that characterise the Hardware Financing framing of how sustainable energy access might be increased in Kenya, in this case, by financing the purchase of solar PV modules and solar home systems (SHSs). The project was to be implemented in three countries simultaneously: Kenya, Morocco and India. Kenya was 'viewed as a true free market for PV products' (IFC 1998, p. 12), presumably therefore making it the ideal context for a market-based intervention (although this clearly begs the question as to why, if the Kenyan market was truly free, would any market-correcting policy intervention be necessary?). With a total investment across the three countries of USD 25 million, the project was expected to have a discernible impact on sales in the world market; specifically, the impact was expected to be about a 5 per cent increase in world PV sales within five years (IFC 1998, p. 14).

A request for proposals was issued in September 1998 (Gunning 2003, p. 85). As the terms of lending were a leverage of 1:1 and a minimum PVMTI investment of USD 0.5 million, companies in Kenya were forced to come together as consortiums because no single company could risk such an amount of money (Bresson 2001, p. 5; Ngigi 2008, interview). One of the first consortiums to submit a proposal involved the Cooperative Bank of Kenya (CBK) together with battery manufacturer Chloride Exide and EAA. This received 'first-track' status, meaning that it was acceptable in principle and ready for implementation (Ngigi 2008, interview). However, the IFC had issues with investing in CBK because of their non-performing assets and decided the proposal was not bankable. Soon after this, according to Ngigi (2008, interview), disparaging articles began appearing in the local media and EAA became one of PVMTI's biggest critics. Certainly by 2001, there was evident disquiet and impatience expressed in the SolarNet[1] newsletter by some actors (de Bakker 2001, pp. 4–5; Bresson 2001, pp. 5–6; Muchiri 2001, p. 4).

Other proposals were received (Hankins and van der Plas 2000; Ngigi 2008, interview), and a long process of negotiations ensued – negotiations between the consortiums and the IFC and, when these failed to produce deals, local financial institutions were persuaded to engage with the project, these deals collapsing after more protracted negotiations (Ngigi 2008, interview). Eventually, it appeared that most of the available finance would finally be disbursed. Three deals were agreed: one with Barclays Bank Kenya, one with Equity Building Society and one with Muramati Tea Growers Savings and Credit Cooperative (SACCO) (Hankins and van der Plas 2000, p. 29). But these fell apart for various reasons, and the disquiet among stakeholders mentioned above turned to resentment.

Byrne (2011) identifies a range of factors that contributed to the failure to broker finance deals for SHSs through PVMTI. These include the following:

1. The minimum deal size was too large for the Kenyan market. Minimum investment was USD 0.5 million from a local consortium, to be matched by PVMTI. No local suppliers were able to mobilise this level of investment on their own.
2. There was misalignment between the IFC and local banking rules, making it impossible for either party to finalise deals.
3. The transaction costs were too high for mainstream' banks, despite some interest in bundling deals for on-lending to micro-finance institutions (MFIs). The deal flows were too small compared with the costs of managing them.

This failure stimulated some actors to begin discussing ways in which PVMTI might be changed in order to provide some tangible benefit to the market (van der Vleuten 2008, interview). These actors approached PVMTI in 2003 requesting help with capacity-building (Magambo 2006, p. 1). In 2004, PVMTI went through a restructuring (IFC 2007b, p. 42). As a result of meetings with PV actors in Kenya and the frustrations felt within the PVMTI hierarchy itself (Ngigi 2008, interview), together with the evidence for training and quality needs (Jacobson 2002a; Jacobson 2002b) and the availability of some technical assistance[2] grant money, PVMTI began a capacity-building project in Kenya in 2006 (IFC 2007b, p. 42; PVMTI 2009). A grant of USD 350,000, together with 'in-kind contributions and co-financing' of USD 115,000, was used to support the Kenya Renewable Energy Association (KEREA), the development of a PV curriculum, PV training courses, the production of three manuals (user, vendor and installer manuals) and a quality assurance programme (Magambo 2006; IFC 2007b, p. 42; Nyaga 2007, interview; PVMTI 2009). PVMTI was then extended to 2011, and local actors began to take a more favourable view of the project (Ngigi 2008, interview).

These latter activities closely resemble various aspects of the capability-building activities noted in previous chapters as being critical to building well-functioning socio-technical innovation systems. Eventually, then, this adjustment of PVMTI's approach resulted in what looked like valuable contributions to fostering the systemic capabilities that provided the bedrock upon which the Kenyan solar PV market developed. But, in terms of the project's main goal – to make a discernible impact on the Kenyan PV market by financing hardware purchases – it was a failure, having helped to finance only 170 SHSs (IFC 2007b, p. 42) in a country that is estimated to have had around 200,000 installed SHSs in 2005, rising to an estimated

320,000 in 2010 (Ondraczek 2013). Similar critiques have been levelled at PVMTI's Hardware Financing-based approach in other countries, such as India (Haum 2012). Hardly a market transformation then.

Socio-Technical Innovation System Building: Lighting Africa

Let us now compare the case of PVMTI to the approach of Lighting Africa, an approach that we have already argued looks like a policy intervention resembling Socio-Technical Innovation System Building. Curiously, this initiative, as with PVMTI, was also funded by the IFC. As already described in Chapter 5, Lighting Africa was launched in September 2007 with a global call for project proposals aimed at developing new lighting products and delivery models for Africa's large, unelectrified, rural off-grid lighting market. The hope was that recent advances in performance of key technologies – especially light-emitting diodes (LEDs) – could be harnessed to provide cheaper and better lighting for the so-called bottom of the pyramid (BOP). Grants of up to USD 200,000 were available for each successful proposal, and 16 were selected from the more than 400 proposals received, four of them to be implemented in Kenya (Lighting Africa 2008c, p. 7). Since then, Lighting Africa conferences have been held in 2010 (Nairobi) and 2012 (Dakar), during which awards were given for a selection of 'outstanding' lighting products on the market. The most recent conference took place in Dubai in October 2015, although it is unclear whether any awards were given this time.

Soon after its launch in 2007, the Lighting Africa programme began implementing a wider range of activities than the call for project proposals. These activities included market research in several countries, product testing and the development of quality assurance methodologies, identification of financing needs throughout the value chain, knowledge-sharing and self-evaluation and moves to identify policy constraints by researching the policy environments in several countries (Lighting Africa 2009). For Kenya, by the end of 2008, the programme had already provided highly detailed qualitative and quantitative market assessments (Lighting Africa 2008a; Lighting Africa 2008b). And much more research followed, including on products available in Kenya, product-testing and a review of the policy environment and policy actors (see the Lighting Africa website[3] for these reports).

Active interventions in Kenya began in 2009 and, over the next few years – up to the official completion of its pilot phase in mid-2013 – the programme engaged in a number of other activities. These interventions included an aggressive and roaming awareness-raising campaign, quality assurance labelling of products, the setting-up of a product quality testing facility, training of technicians, capacity-building for business development and for finance institutions, lobbying of policy-makers on regulations and building networks of actors to encourage the flow of information. The systemic nature of these capability-building activities undertaken by Lighting Africa, together with their direct engagement with potential technology users and attention to the kinds of technologies that might best suit local needs, is much closer to a Socio-Technical Innovation System Building approach than to either Hardware Financing or Private Sector Entrepreneurship interventions. While it is difficult to determine the extent to which outcomes can be attributed directly to these efforts, the programme does make a series of claims about its impact across Africa (see Table 6.1). And a recent updated survey in three towns in Kenya tends to support the notion that the market for pico-solar off-grid lighting products has expanded rapidly since 2009. Specifically, the survey report claims that cumulative sales of quality-assured pico-solar products since 2009 had reached about 2.3 million in Kenya by June 2015 (Turman-Bryant et al. 2015, p. 6). This suggests that the Lighting Africa approach has been far more transformative than PVMTI could ever claim to have been.

PVMTI versus Lighting Africa

The SHS market in Kenya is now worth about USD 6 million annually, and more than 320,000 SHSs have been installed (Ondraczek 2013). We do not have figures for the value of the pico-solar market but, as reported above, the number of products sold is significant. This is still a relatively small fraction of the population with access to small quantities of electricity from PV. Nevertheless, as argued in Chapter 5, the case does represent one of the best examples in Africa of anything resembling a transformation in relation to sustainable energy access.

Little – if any – of the success can be attributed to the Hardware Financing intervention PVMTI. But the IFC may have learned important lessons from the PVMTI experience that informed the design of the Lighting Africa programme – whether or not this is the case is an empirical question in need of further research.

TABLE 6.1 Lighting Africa claimed impacts as of December 2014

Quantity	Impact
35,000,000	People across Africa with improved energy access due to modern solar lighting products
14,380,000	People in Africa whose basic lighting needs are being met by modern solar lighting products
7,500,000	Quality solar lighting products sold through local distributorships in Africa
700,000	Tons of GHGs avoided in Africa
31	% growth in sales of quality-assured solar lights between July and December 2014
4.8	% of Africa's un-electrified now using solar lights up from less than 1% in 2009
41	Manufacturers whose products have passed the Lighting Global Quality Standards
25	Countries where quality-verified products are on sale in Africa

Source: Selected impacts reported in Lighting Africa (2014).

The Lighting Africa programme has taken a more systemic approach to developing the market, albeit a different segment of the market than the target of PVMTI. It has focussed more on building capabilities throughout the value chain, building actor networks and influencing policy and other institutions. These are all supply-side activities. However, crucially, from the perspective of this book, Lighting Africa also built a detailed understanding of the electricity needs and desires of poor people, using this to inform its awareness-raising efforts and, thereby, helping to create demand for solar lighting alternatives to kerosene.

So, one clear lesson from the evolution of the Kenya PV market is that its success – whether for SHSs or pico-solar lighting products – is explained by a combination of interventions that has addressed several dimensions of a nascent socio-technical innovation system. An interesting aspect of these interventions has been the attempt to understand the detail of consumer preferences and constraints. This has enabled much better designs of SHSs and lighting products that address the context-specific nature of electricity services in rural areas. And the results suggest that governance of inclusive energy transitions would be improved by taking a systemic approach and that, in this approach, working more closely with consumers of energy services to understand their needs more deeply would raise the chances of providing more appropriate solutions.

Private Sector Entrepreneurship: Climate Innovation Centres (CICs)

Having explored the relative success of Lighting Africa's Socio-Technical Innovation System Building approach next to PVMTI's Hardware Financing approach, we now spend some time examining a current policy intervention that typifies the Private Sector Entrepreneurship framing. This is the example of the CICs introduced briefly in Chapter 1. Although, as the analysis below reveals, the CICs are to some extent already engaged in activities that could move them beyond their predominant focus on private sector entrepreneurship, there is significant potential to do much more. While it is too early to judge the success of the Kenyan CIC and, furthermore, we lack the empirical evidence to do so (another area warranting further research), it is possible from publicly available documentation to examine the way in which the CIC approach is framed. It is also possible to look at the outcomes reported to date. It is on the basis of such *ex-ante* analysis, building on the lessons from the comparison of Lighting Africa and PVMTI above and the theoretical insights from this book, that we analyse the CIC approach below.

CIC framing and approach

The CICs are being implemented as the 'flagship initiative' (infoDev 2015, p. 4) of infoDev's (the World Bank's innovation arm) Climate Technology Programme (CTP), in collaboration with DFID and Danida (the UK and Danish overseas development agencies). At the time of writing, CICs have been launched – or have business plans and are in the process of being launched – in seven locations: Kenya, India, Ethiopia, South Africa, Morocco, Vietnam and the Caribbean. The CICs' focus is very much on financing local entrepreneurship around climate technologies via 'a tailored suite of financing and services that support domestic SMEs' (infoDev 2015, p. 8).

The overall framing and associated narratives with which the CTP operates are well captured by the following statements taken from the CTP brochure (infoDev 2015, pp. 4, 8):

> The CTP supports the private sector in developing countries – targeting SMEs and entrepreneurs – to innovate novel technologies and business models to address local climate challenges.
>
> The CICs build local capacity and address barriers to innovation by offering a tailored suite of financing and services that support domestic SMEs. With the CIC's assistance, innovative enterprises become more competitively and profitably involved in booming local and international cleantech markets – creating jobs and leading to economic growth.

The document goes on to describe how knowledge bridging is also being attempted across different countries' CICs, providing a means of sharing and accessing knowledge internationally, including policy insights.

The quotes above demonstrate how the CTP and CICs frame small and medium-sized enterprises (SMEs) and entrepreneurs as the key actors in addressing local climate challenges, including energy access, an issue mentioned several times in the document as something the CICs will positively impact. The CTP website claims one of its impacts will be to 'provide energy access to over 28 million people ...',[4] and the brochure claims that the programme will provide 'clean energy to an additional 35,000 homes in South Africa' and '1,400MW of off-grid renewable energy access' (infoDev 2015, p. 5). It should be noted that energy access is just one planned impact of the programme among a range of others, including:

- tapping into, in unspecified ways, the 'vast untapped resource' of women and girls' contribution to 'climate sectors' (infoDev 2015, p. 7);
- 'allowing 3,000,000 people including women and girls to be less vulnerable to the effects of climate change' (infoDev 2015, p. 5);
- 'allowing access to clean water for 90,000 Ethiopian households' (infoDev 2015, p. 5).

This casts the claimed benefits of funding private sector entrepreneurship in relation to climate technologies very widely, intersecting with multiple other development priorities, including gender, climate adaptation and resilience, clean water and sanitation, as well as sustainable energy access. These claims are then followed with a narrative that depicts supporting these actors with venture financing as the key to plugging both local innovators and developing countries 'more profitably' into 'booming local and international cleantech markets' with resulting benefits in terms of jobs and economic growth (infoDev 2015).

The implication is that this kind of financial and other business support to private sector entrepreneurs is the means through which transformative impacts can be achieved in sustainable energy access as well as a raft of other climate-related areas. Bearing in mind the much broader suite of both actors and activities that previous chapters have demonstrated are necessary for well-functioning socio-technical innovation systems, we might readily conclude that this kind of Private Sector Entrepreneurship approach is likely to meet with limited success. It certainly does not look like the kind of systemic intervention that we have so far argued is necessary to transform sustainable energy access. When we look at the list of activities that the CICs are meant to conduct, however, it becomes clear that the CICs' activities are, in fact, intended to be broader than a single focus on financing private sector entrepreneurs.

The Kenyan CIC from a Socio-Technical Innovation System perspective

Each of the national-level CICs is launched with its own business plan developed from a consultative process that engages with a range of private, NGO and public sector stakeholders. The Kenyan CIC's business plan (summarised on its website) states the following mission, intended impact and goals:[5]

Our Mission

Our Mission is to provide an integrated set of financing, venture acceleration services, market development and networking activities that builds the quality and quantity of Kenyan entrepreneurs and startups delivering innovative climate and clean energy solutions to local and international markets.

Impact

The CIC is here to deliver a mix of economic, environmental and social results, including job creation, reduction of carbon dioxide emission, greater climate resiliency, and access to clean energy, safe drinking water, better sanitation and strengthened technology transfer and local innovation capacity.

Our Goals

- Providing flexible financing mechanisms that support entrepreneurs and new ventures at varying levels of innovation and scale.
- Building innovation capacity through the delivery of advice, assistance and training products.
- Enabling collaboration and developing policies that support an innovation ecosystem in East Africa.
- Identifying and unlocking new opportunities through access to information and market intelligence.

- Providing access to facilities that support business development through co-working and networking space and technical development for rapid design, adaptation, proto-typing, testing and manufacturing.

While the mission is firmly focussed on Kenyan entrepreneurs and start-ups, the goals of the CIC make for interesting reading in the context of this book. The second goal speaks of 'building innovation capacity', the third of supporting 'an innovation ecosystem in East Africa', and the fourth mentions 'access to information and market intelligence'. Both 'building innovation capacity' and supporting 'an innovation ecosystem' are clearly linked to aspects of the Innovation Studies theory introduced in Chapter 2, and facilitating 'access to market intelligence' was one of the key contributions of donors highlighted in our analysis in Chapter 4 and Chapter 5. Clearly, then, at least at the level of the Kenyan CIC's aims and ambitions, a more systemic approach to transforming sustainable energy access, or markets for other climate-relevant technologies in Kenya, is an idea with which the CIC is sympathetic.

When we look at the various 'global activities' that the CTP brochure (infoDev 2015) sets out, which are designed to support entrepreneurs and SMEs and to network CICs across the different constituent countries where they are being rolled out, we see additional activities that resemble more systemic interventions. These activities (detailed in the brochure and summarised on the CTP website) include:[6]

- *CIC Design and Implementation*: CIC LAUNCH will lead scoping, design, resource mobilisation and implementation activities to meet client country demand for CICs.
- *Global Financing*: The IGNITE Fund will mobilise and syndicate global funding for high-impact climate technologies and offer deal-flow to public and private investors eager to support promising climate ventures in developing countries.
- *Evidence-based Analysis*: Climate TRACK will actively package lessons from individual CICs and provide cutting-edge analytical products and policy toolkits on supporting private sector innovators in developing countries.
- *Connecting Markets*: Market CONNECT will provide software, web-enabled services and networking technology to build interconnectivity between CICs and link promising companies with global partners and expertise.
- *Measurement Tools*: Impact Xchange will provide each CIC with a web-based Impact Monitoring System (IMS) to track results and impacts in real-time.

The ideas of lesson-sharing and learning in relation to Climate TRACK and the networking activities under Market CONNECT are certainly, on the face of it, activities that intersect with some of the key elements of the Socio-Technical Innovation System Building highlighted in previous chapters.

As emphasised above, without more in-depth empirical research, it is impossible to get an accurate picture of the extent to which actual activities on the ground are delivering against these stated aims, which pertain to pursuing more systemic interventions. From the perspective of the public information available on the Kenyan CIC website, the actual activities being pursued look, on the face of it, to fall well short of the stated aims above regarding building what they call innovation ecosystems. Activities are grouped around four categories:[7]

1. *Business Acceleration*: Kenya CIC provides diverse services to support entrepreneurs in cleantech to grow their ideas.
2. *Financing*: CIC aims at facilitating flexible access to finance to fund clean technology businesses through their growth cycle.
3. *Market Development*: The CIC provide market intelligence products including the market opportunity for various clean technologies.
4. *Matchmaking*: The KCIC provides a one-stop matchmaking service that offers a range of quality services.

The market development activity sounds like it might deliver something of interest in the context of the activities described in previous chapters regarding donors investing and making publicly available market research. Drilling down under the market development activity, the website claims that the CICs will provide:[8]

- *Market Information*: The KCIC provides market intelligence products including the market opportunity for various clean technologies, current market penetration, and information on ideal price points for large-scale consumer adoption. We also provide information on competing solutions in the market.
- *Sector Trends*: The KCIC provides technical information on a regular basis to enable technology development in line with market needs. The CIC is developing a database that will provide entrepreneurs with information on cleanTech component sourcing as well as providing consumers with clean technology options to meets their needs.
- *Technical Information*: The KCIC provides a database of financial support from various sources available in Kenya to provide entrepreneurs and consumers with information on financing options. The CIC also provides access to the latest information, market research and trends for various technologies, enabling technology to be developed in-line with market needs.

But there is nothing on the website that resembles this kind of information. The only relevant material in the document repository is a report and policy brief on the regulatory environment for solar energy in Kenya. Perhaps the intention is that this kind of market development information will be provided on a bespoke basis to individual CIC 'clients' (as the website refers to the entrepreneurs and the SMEs the CIC supports). But such private provision of information is not in line with the publicly available market research funded by donors during the development of the solar PV market. It would therefore be unlikely to underpin any transformative change, benefiting, as it would, only the individual private 'clients' in receipt of such CIC-funded research.

The website showcases an impressive range of example clients across the three sectors on which the Kenyan CIC focusses (renewable energy, water management and agri-business). Drilling down into the details of the services that each client is receiving, however, it seems that the main focus is on providing support for accessing finance, developing business plans and other (often unspecified) business development support. Interestingly, several of the renewable energy-related examples include provision of support to test and improve technology hardware, presumably by funding access to R&D and testing facilities through one of the Kenyan CIC partner organisations. None of the activities that seem to currently be underway in the Kenyan CIC, therefore, seem to be focussed on anything much beyond assisting with accessing finance or other mainstream business incubation activities, together with some support for technology hardware development.

Notwithstanding the caveat of further empirical research being required to triangulate the analysis presented here, it does seem as though the CIC's current activities are squarely located within a two-dimensional technology-finance approach. This mirrors the dominant two-dimensional focus of the literature on energy access in Sub-Saharan Africa that we critiqued in Chapter 1. The predominant emphasis on financing SMEs and entrepreneurs also positions the CICs as typical of a Private Sector Entrepreneurship framing of potential solutions to sustainable energy access. It would not be unreasonable, therefore, bearing in mind the alternative perspective proposed in this book, to question the extent to which the CICs are likely to deliver the kinds of transformative long-term impacts that their brochure claims they will.

However, we do not wish to argue that the CTP and constituent CICs are negative initiatives. Funding or financing entrepreneurs to engage in innovative activities that engage with any of the climate-relevant sectors the CICs support must have *potential* social, economic and environmental benefits. As we were at pains to emphasise in the Introduction to this book, our argument is certainly not that hardware financing or private sector entrepreneurship are unimportant. They aren't. Furthermore, it seems clear both from the evidence available in the public realm, and the authors' conversations with stakeholders in Kenya, that the CICs are having a positive impact in terms of convening diverse stakeholders around the issue of climate technology and innovation. This, at least, is providing a new platform on which these stakeholders can engage with one another, develop new networks, or cement existing ones. These are all critical to the development of well-functioning socio-technical innovation systems.

But the question remains, no matter how successful the SMEs and entrepreneurs supported by the CICs become: is this enough to transform sustainable energy access or, for that matter, the CICs' other focal sectors? To what extent will the new technologies and business models supported by the CICs lead to wider impacts across Kenya? Will they remain isolated examples of SME success stories, or will they connect with other stakeholders, building momentum and improving technological capabilities across the country, enhancing its innovation system? It seems quite an ambitious assumption to think that the creation of a number of new companies – the brochure lists a target of 2,500 new companies across all seven countries where CICs are being established – will be enough to transform energy access, or technology availability and adoption in other climate-relevant sectors. In what way does the CIC approach encourage innovation that engages with the social practices of technology users? To what extent will they be able to compete with existing energy interests (or the interests of mainstream, non-sustainable agricultural and water management actors) and influence dominant socio-technical regimes? Is the expectation that this will all happen of its own accord, or is something more deliberate and proactive required to connect the dots and lift the CICs beyond their core focus on financing private sector entrepreneurship and business incubation?

These are all empirical questions. Some are perhaps answerable now with proper empirical research efforts, some perhaps answerable in time, once we reach a point where we are able to look back at the history of the CICs and what they were able to achieve. But it would be a shame, particularly in light of the historical evidence analysed in this book, not to at least attempt some level of a priori analysis and recommendations for policy and practice that might catalyse transformative change.

Of course, Socio-Technical Innovation System Building could be integrated as part of the CICs' activities under an extended remit. This may be something that infoDev, DFID, Danida, and national governments and local partners in the CICs might wish to consider in future. CICs are, after all, likely to represent important networks of climate technology-relevant individuals and organisations across the public, private and NGO sectors, and provide excellent potential routes for identifying and engaging with key actors. Indeed, as discussed below, innovation system building was an explicit intention of the originators of the idea that eventually became the CICs (Sagar et al. 2009).

Practical approaches to Socio-Technical Innovation System Building

Having conducted a somewhat critical analysis of PVMTI and the CICs, it would be remiss of us not to conclude the analysis in this chapter with some concrete suggestions of how policy could be designed in ways that operationalise a Socio-Technical Innovation System Building approach. In the final section of this chapter, we therefore describe an approach that builds on Ockwell and Byrne (2015), which elaborates proposals that responded to a new interest within the United Nations Framework Convention on Climate Change (UNFCCC) in achieving technology transfer by means of strengthened national innovation systems. While the paper dealt specifically with the context of the UNFCCC, here we describe how this policy approach is equally relevant to efforts to transform access to sustainable energy in developing countries. We begin by outlining the key policy goals that are implied by the theoretical framework developed in Chapter 2 and Chapter 3, and the empirical analysis given in Chapter 4 and Chapter 5. We then proceed to outline two linked policy proposals through which these policy goals might be achieved.

Overarching policy goals

The overall goal of policy must be to build functioning socio-technical innovation systems that augment the transfer, development and diffusion of sustainable energy technologies and related practices in developing countries, enhancing technological capabilities through a range of targeted interventions. These must be inclusive in their approach – attending to the self-defined needs of those countries and different groups within – if sustainable energy technology adoption is to be widespread and underpin pro-poor sustainable development pathways. Material presented in this book provides some clues as to what such an inclusive approach might be. The various interventions described in Chapter 4 and Chapter 5 – where they have achieved some measure of success – were designed and implemented on the basis of careful and context-specific understandings of the needs in the market and of users. Notable in this regard is Lighting Africa, which conducted highly detailed studies of the lighting practices and needs of poorer users in Kenya (and elsewhere). This suggests that further gains might be achieved by including users more actively in the design of promising solutions to their needs, rather than merely observing these needs and eliciting users' feedback on products already on the market. The overall desired result is to provide protective spaces in which sustainable energy technologies and practices can be fostered, thus promoting their adoption, adaptation and further innovation.

In order to achieve this, we suggest the following overarching policy goals should orient interventions. However, it is important to note that interventions to build socio-technical innovation systems are deeply interdependent. They are therefore best implemented together in systemic fashion rather than separately. We conclude this subsection with a list that articulates a range of specific policy interventions that could be pursued in order to fulfil each goal. In subsequent sections, we go into detail on how interventions through policy and practice could deliver such interventions.

Goal 1: Build networks of diverse stakeholders

Efforts are required to link diverse arrays of stakeholders, from technology importers and suppliers through to policy-makers and technology users. Such networks enable the flow of knowledge among stakeholders, each of whom can bring different resources, experiences and perspectives to bear on problem-framing and problem-solving activities. They can also become a fundamental element of socio-technical innovation systems by establishing the linkage component of capabilities. But these linkages must be strong and meaningful. In order to achieve this, stakeholders need to work proactively together in projects, programmes and other interventions. In doing so, they are more likely to build mutual trust and understanding, as well as identify strengths and weaknesses in local technological capabilities. Simultaneously, by pursuing such activities, new technological capabilities can be built, including the development of relevant knowledge and skills.

Goal 2: Foster and share learning

Learning is critical to the development of technological capabilities and functioning innovation systems, and the resulting successful markets for climate technologies that these can support. A key role for policy lies in commissioning research – whether market research, academic analysis, monitoring and evaluation, baseline studies, R&D and so on – and making sure the results are publicly available. Because contexts evolve in unpredictable ways, incremental innovation supported by reflexive analysis offers a practical strategy to shape sustainable development pathways. Research at all levels from local to international, and from different perspectives, can provide crucial information to help realise such reflexive change. The public availability of such information can play a fundamental role in reducing perceived risks among both potential investors and technology users, as well as enhance the transparency of policy processes. This facilitates clear and evolving understandings of things such as user needs and preferences, appropriate hardware components relative perfor-mance of different technology brands approaches that have met with success factors that contributed to difficulties or failures and how to overcome these training and education needs and so on. The learning and experience that result can feed into future projects and programmes, whether publicly or privately funded.

Goal 3: Promote the development of shared visions

Linked to the need to build meaningful networks and foster learning, there is the need to create shared visions of what pro-poor sustainable energy access – and sustainable development more broadly – look like in particular contexts, and what roles different sustainable energy technologies play in those contexts. This is not simply a top-down effort in which sustainable energy technology solutions are chosen and then stakeholders are persuaded of their merit through dissemination and awareness-raising activities. As everyone is affected by both sustainable devel-opment issues and efforts to address them, consensus-building around sustainable development, including sustainable energy access, is critical. Learning from research and experience provides an essential component for constructive debate and is itself enhanced by the flow of knowledge through diverse stakeholder networks. By fostering understandings of what sustainable energy technologies can and cannot provide, how they work and the ways others have benefited from them, visions can develop around informed understandings of different technological options. It also affords opportunities for users to provide feedback on both their self-defined needs and their experiences (good and bad) with different technologies. As a result, shared visions develop among technology users, suppliers and other stakeholders relating to how and in what way sustainable energy technologies can underpin different development pathways. This simultaneously provides vital user feedback into both technology design and the configurations and brands that vendors and suppliers provide, with attendant implications for potential market size and profitability.

Goal 4: Support diverse experimentation

Again linked to learning, funding is needed to provide protected spaces for experimentation with promising sustainable energy technologies, practices and policies. Stakeholders throughout the supply chain need to gain experience of technologies and learn what works and what does not within specific contexts (across different countries, regions, villages, technologies, social practices, political contexts, etc.). Experimentation can target a range of different aspects. It might, for example, include supporting new multi-stakeholder projects that test and develop ideas. These could relate to new technical configurations, new hardware, new practices around existing technologies, new consumption and production practices that could improve the benefits accrued by users and so on. Experiments might also focus on mutually supportive interventions that link different stakeholders across markets, thereby building supply chains and fostering new market opportunities where potential market players lack awareness of each other or potential market opportunities they might target. Interventions could also experiment with working 'upwards' through value chains, building on existing markets to develop progressively higher-value segments, adding value to existing sectors and fostering increasing economic returns from sustainable energy technology initiatives across developing countries.

Specific policies and interventions for delivering against these overarching goals

Below is a (non-exhaustive) list of specific policies and interventions that could deliver against these overarching goals and contribute to Socio-Technical Innovation System Building.

Goal 1: Network building

- Link diverse stakeholders nationally.
- Link diverse stakeholders internationally.
- Link diverse stakeholders locally.
- Link diverse stakeholders across markets.
- Link diverse stakeholders across sectors (private, public, NGOs, research, etc.).
- Link supply-side actors (e.g. supply chain, policy, NGOs, etc.) with technology users.
- Link national government with technical experts.
- Link national firms with international firms.

Goal 2: Learning

- Commission market research.
- Commission research into technology user needs and preferences.

- Commission research into technology performance.
- Commission research into education and training needs.
- Monitor and evaluate projects and programmes.
- Conduct baseline studies.
- Conduct comparative research across local, national, international scales that addresses the various research foci above.
- Make results of research and monitoring and evaluation publicly available.
- Create spaces for stakeholders to reflect on research and experiences.
- Provide training for firms.
- Provide training for suppliers and installers.
- Provide training for technology users, villages, households.
- Advise on and develop technology certification schemes.
- Advise on education and training needs (up to and including postgraduate training).

Goal 3: Foster shared visions

- Convene consensus-building events with different national stakeholder groups.
- Convene scenario-building events to discuss alternative development pathways that different sustainable energy technologies might contribute to or constrain.
- Facilitate opportunities for different stakeholders to feed back into the technology design and configuration process.

Goal 4: Provide protected spaces for experimentation

- Encourage and incentivise treatment of 'failures' as valuable points for learning.
- Commission projects as experiments (examples of potential foci for experimentation are provided below).
- Experiment with technological hardware.
- Experiment with policies.
- Experiment with social practices in relation to sustainable energy technologies.
- Experiment with new stakeholder configurations.
- Experiment with production processes.
- Experiment with linking stakeholders across markets to create new market opportunities and market awareness.
- Experiment with value-adding experiments, working upwards through supply chains.

Existing international policy mechanisms

While working towards the overarching goals above, it is essential that policies designed to nurture socio-technical innovation systems are implemented in a way that recognises and builds on existing relevant policy mechanisms and institutions. Designing effective policy also requires an understanding of what these existing initiatives are doing that is of relevance to nurturing socio-technical innovation systems and where there are gaps that need to be filled. Here we review two core areas of relevant policy efforts: (1) the UNFCCC's Climate Technology Centre and Network (CTCN); and (2) four parallel climate technology centre and network initiatives currently being funded by the Global Environment Facility (GEF). We then provide a visual overview of the coverage of these existing programmes, together with the CICs, against the overarching policy goals articulated above.

It should be noted that a range of other institutions (e.g. the International Renewable Energy Agency, IRENA), policies, mechanisms (e.g. the Clean Development Mechanism, CDM) and centre-based models (e.g. Innovación Chile and CGIAR, the Collaborative Group for International Agricultural Research) also exist and deserve consideration when implementing the recommendations in this chapter. It is, however, beyond the scope and space available here to provide a full review of all relevant initiatives. We have therefore opted instead to focus on the most relevant emerging initiatives.

The Climate Technology Centre and Network (CTCN)

In the context of actions under the Convention (the UNFCCC), one of the most relevant institutions is the Climate Technology Centre and Network[9] (CTCN), the operational arm of the UNFCCC's Technology Mechanism under the strategic guidance of its own Technology Executive Committee (TEC). As its name suggests, the CTCN is structured around a core climate technology centre that coordinates a broader network. The Centre is hosted and managed by the UN Environment Programme (UNEP) in collaboration with the UN Industrial Development Organisation (UNIDO) and support from 11 centres of excellence located in developing and industrialised countries.

The CTCN's *Network* refers to a range of technical experts and centres of excellence who have expertise that might be matched against requests for technical assistance from countries. Requests from countries come from national designated entities (NDEs). NDEs[10] (usually government ministries or agencies) are granted responsibility by Parties to the Convention to manage national technology-related requests to the CTCN. These requests are coordinated by the Centre, which responds itself to some, while others are farmed out to relevant experts in the Network. This NDE-instigated approach attempts to facilitate a process that is demand-driven by Parties. There are three core services offered by the CTCN (see CTCN 2014 for a detailed description of these services):

1.   Provide technical assistance to developing countries to enhance transfer of climate technologies.
2.   Provide and share information and knowledge on climate technologies.
3.   Foster collaboration and networking of various stakeholders on climate technologies.

The first core service follows requests from NDEs, while the other two services can be initiated by the CTCN or other stakeholders as and when common needs are identified.

From the perspective of building socio-technical innovation systems, there are several key points to note with regard to the CTCN:

1. The Network is not an in-country network of actors of relevance to different (existing or emerging) climate technologies as prescribed in the list of overarching goals above.
2. There is nothing, in theory, stopping Parties requesting, via NDEs, support from the CTCN in advising on and instigating the kind of Socio-Technical Innovation System Building policies detailed in our list above.
3. NDEs are usually government institutions – not locally nested, climate technology-specific institutions.
4. At present, the CTCN's activities do not explicitly recognize the need to nurture socio-technical innovation systems as a key part of the technology transfer, development and diffusion process – although elements of innovation system building are implicit within two of the CTCN's core services: those that focus on information and knowledge sharing, and fostering collaboration and networking between stakeholders.
5. The recognition of knowledge-sharing, networking and the emphasis on capacity-building elaborated in the operating manual for NDEs suggests significant potential for the CTCN to coordinate its attempts to achieve a stronger focus on Socio-Technical Innovation System Building. However, this would require more explicit attention to, and understanding of, Socio-Technical innovation System Building, and processes for strengthening them to be integrated into the CTCN's approach
6. The TEC has recently initiated a work stream focusing on National Systems of Innovation as a means for strengthening efforts under the UNFCCC to develop and transfer climate technologies. This suggests considerable potential to engage productively with efforts under the CTCN and UNFCCC more broadly around implementing the policy proposals outlined further below.

Global Environment Facility-funded initiatives

The other key initiatives of note here are those being implemented by the GEF under its Long-Term Program on Technology Transfer. These include:[11]

1. The project 'Pilot Asia-Pacific Climate Technology Network and Finance Center', which is being implemented with the Asian Development Bank (ADB) and UNEP.
2. The project 'Finance and Technology Transfer Centre for Climate Change' by the European Bank for Reconstruction and Development (EBRD).

3. The project 'Pilot African Climate Technology Finance Center and Network' by the African Development Bank (AfDB) (which includes regional partners that are part of the CTCN consortium).

4. The regional project 'Climate Technology Transfer Mechanisms and Networks in Latin America and the Caribbean (LAC)' by the Inter-American Development Bank (IADB), which is currently in preparation, again with regional partners that are part of the CTCN consortium.

As with the CICs, the emphasis of the second of these various initiatives (the EBRD one) is mostly focussed on finance. However, the other three all have elements that pertain to a more networked, capacity-building focus and hence have potential to act as socio-technical innovation system builders. For example, the ADB-led Asia-Pacific initiative (number 1 above) includes aims[12] of facilitating a network of national and regional technology centres, organisations and initiatives; building and strengthening national and regional climate technology centres and centres of excellence; designing, developing and implementing country-driven climate technology transfer policies, programmes, demonstration projects and scale-up strategies. These activities are pursued in parallel to another part of the initiative, that focusses explicitly on finance.

While detailed information is difficult to obtain on number 3 (the AfDB-led African initiative), it seems that, as well as a core finance component, more network and capacity-building activities will be included, with publicity materials released by AfDB suggesting that 'enhancing networking and knowledge dissemination' is seen as the key way the project will 'scale-up deployment of [climate technologies]'.[13] The final one (number 4, IABD) is not yet operational. However, it is very much focussed on network and capacity-building. As well as providing finance, it seeks to 'strengthen existing activities on [environmentally sound technologies] in LAC and aim at the consolidation of long-term collaborative initiatives that are aligned with the objectives and modalities of the Technology Mechanism under UNFCCC'.[14] Planning, assessments and networks are very much at the foreground of the activities proposed under this initiative.

As with the CTCN, however, the extent to which these GEF-funded initiatives support the development of socio-technical innovation systems depends on the extent to which an explicit focus on innovation system building can be mainstreamed across the various activities. The language used is certainly open to a systemic perspective, but achieving real impacts will depend on more deliberate integration of Socio-Technical Innovation System Building activities across the board.

Gaps analysis of existing policy

In order to get an overview of the extent to which the initiatives reviewed above are delivering the kind of policy interventions that would be likely to achieve Socio-Technical Innovation System Building, delivering against the overarching

TABLE 6.2 International policy mechanisms and innovation system building goals

Climate innovation system building goals	CTCN	CIC	ADB	EBRD	AfDB	IADB
Explicit focus on climate innovation system building?	N	N	N	N	N	N
1. Network building						
Linking diverse stakeholders nationally	P	Y	Y	N	P	Y
Linking diverse stakeholders internationally	Y	Y	Y	N	Y	Y
Linking diverse stakeholders locally	P	P	P	N	P	P
Linking diverse stakeholders across markets	P	P	Y	N	P	P
Linking diverse stakeholders across sectors (private/public/NGO/research, etc.)	Y	P	Y	N	Y	P
Linking supply-side actors (e.g. supply chain, policy, NGO, etc., actors) with technology users	P	P	P	N	P	P
Linking national government with technical experts	Y	P	Y	Y	Y	Y
Linking national firms with international firms	Y	P	Y	N	Y	P
2. Learning						
Commission market research	P	N	P	N	P	P
Commission research into technology users' needs and preferences	P	N	P	N	P	P
Commission research into technology performance	P	N	P	N	P	Y
Commission research into education and training needs	P	N	P	N	P	P
Monitoring and evaluation of projects/programmes	P	N	Y	N	P	P
Conduct baseline studies	P	N	P	Y	P	P

Climate innovation system building goals	CTCN	CIC	ADB	EBRD	AfDB	IADB
Conduct comparative research across local/national/international scales that addresses the various research foci above	P	N	P	N	P	P
Make results of research and monitoring and evaluation publicly available	P	N	P	P	P	P
Create spaces for stakeholders to reflect on research and experiences	P	N	P	N	P	Y
Provide training for firms	P	N	P	P	P	P
Provide training for suppliers and installers	P	N	P	P	P	P
Provide training for technology users/villages/households	P	N	P	N	P	P
Advise/develop technology certification schemes	P	N	Y	P	P	Y
Advise on education and training needs (up to and including postgraduate training)	P	N	P	P	P	P
3. Foster shared visions						
Convene consensus-building events with different national stakeholder groups	P	N	P	N	P	P
Convene scenario-building events to discuss development pathways that different climate technologies might contribute to/constrain	P	N	P	N	P	P
Facilitate opportunities for different stakeholders to feed back into the technology design and configuration process	P	N	P	N	P	P
4. Provide protected spaces for experimentation						
Encourage/incentivise treatment of 'failures' as valuable points for learning	P	N	P	P	P	P
Commission projects as experiments (examples of potential foci for experimentation are provided below)	P	N	P	P	P	P
Experiment with technological hardware	P	N	P	P	P	P

Table 6.2 (continued)

Climate innovation system building goals	CTCN	CIC	ADB	EBRD	AfDB	IADB
Experiment with policies	P	N	P	N	P	P
Experiment with social practices in relation to climate technologies	P	N	P	N	P	P
Experiment with new stakeholder configurations	P	N	P	N	P	P
Experiment with production processes	P	N	P	P	P	P
Experiment with linking stakeholders across markets to create new market opportunities and market awareness	P	N	P	P	P	P
Experiment with value-adding experiments working upwards through supply chains	P	N	P	P	P	P

IADB = Climate Technology Transfer Mechanisms and Networks in Latin America and the Caribbean (LAC) – Inter-American Development Bank (IADB) goals articulated above, Table 6.2 provides a graphical overview of the current and potential coverage of each initiative. Table 6.2 also provides a useful overview of the aggregate pattern of coverage across the initiatives. Each initiative is assessed, based on available public documentation, on the extent to which it: (1) explicitly includes activities akin to the policy options under each goal within its existing remit and structure (the Ys for 'yes'); (2) has potential to deliver against a policy option within (or with incremental adjustments to) its existing remit and institutional structure (the Ps for 'possible'); and (3) requires significant revisions to remit and institutional structure in order to deliver against a goal (the Ns for 'no'). The initial row also indicates whether Socio-Technical Innovation System Building is an explicit goal of each initiative.

Several key observations can be made from Table 6.2:

1. Most initiatives have potential within, or via incremental adjustments to, their existing remit and structure to extend their activities to include Socio-Technical Innovation System Building activities.
2. At present, however, there is limited focus on activities that would nurture socio-technical innovation systems in developing countries.
3. The most coverage exists in the area of network building. However, even this coverage is patchy, with most initiatives focussing on high-level national or, more commonly, international networking activities, or linking national entities with international technical experts. Many of the essential networking activities that are necessary to build socio-technical innovation systems in ways that will result in sustained pro-poor socio-technical change are generally not addressed (e.g. linking with technology users or fostering local networks along supply chains).
4. Learning receives a small amount of patchy coverage across the initiatives.
5. Fostering shared visions and providing protective spaces for experimentation are not covered at all at present.

Two linked policy proposals

Our Socio-Technical Innovation System Building approach can be implemented using two distinct but linked proposals, as summarised here and described in more detail below:

1. *Creation of Sustainable Energy Access Relevant Innovation-system Builders*[15] *(SEA-RIBs)*. This involves the creation of specific institutions (preferably based in existing organisations) in different countries that are focussed on building socio-technical innovation systems around sustainable energy technologies (SEA-RIBs). These institutions would then work to coordinate efforts to implement Proposal 2 below, as well as engaging in broader functions (described in more detail below). This approach could be pursued as an initiative under SE4All. It could also be integrated with an extension of the UNFCCC architecture (specifically with the CTCN) and/or link with various

centre-based initiatives under the GEF (see Ockwell and Byrne 2015 for more detail on the nature of such integration). It is also feasible for it to be pursued bi- or unilaterally through donor, individual government or NGO-driven activities. As emphasised below, however, such centres would be likely to have greater impact if networked across different countries and regions (e.g. via the CTCN).

2. *Using projects and programmes to build socio-technical innovation systems.* This proposal focusses on the possibilities of designing project- and programme-based investments in sustainable energy access to maximise impacts on building socio-technical innovation systems. It can be pursued by any actors engaged in work on sustainable energy access. Again, it is likely to have far greater impact if actors pursuing such approaches coordinate via new or existing networks.

Proposal 1 could be adopted in a way that is integrated with Proposal 2, working to ensure that as many project-based interventions as possible are pursued in ways that maximise opportunities for Socio-Technical Innovation System Building. Proposal 2, on the other hand, could be pursued in isolation from Proposal 1. For actors involved in attempts to improve sustainable energy access, it could represent a strategic commitment to maximising the impact of their activities on Socio-Technical Innovation System Building. Clearly, however, the transformative impact of activities pertaining to Proposal 2 is likely to be far higher if pursued as part of nationally and internationally coordinated efforts via the kinds of institutions suggested in Proposal 1. We describe each proposal in more depth below.

Proposal 1: Creation of Sustainable Energy Access Relevant Innovation-system Builders (SEA-RIBs)

As the analysis in Chapter 4 and Chapter 5 has demonstrated, the socio-technical innovation system that emerged around the Kenyan PV market was, in important ways, developed via the targeted, long-term efforts of specific actors, or 'champions', acting as socio-technical innovation system builders. This proposal is therefore intended to create institutions that are able to take on such a system-building role. There are examples of similar roles having been taken on in the past, such as the CGIAR, Innovación Chile and the UK's Carbon Trust – all of which represent nationally situated, long-term institutional presences that pursue approaches that are sensitive to the needs and contexts of the people and organisations with whom they engage. The proposal here is to learn from the successes of such nationally focussed, strategic initiatives and bolster them with insights from the kind of in-depth historical research on what has worked for sustainable energy access in the past, such as the empirical evidence presented in this book.

SEA-RIBs would play a strategic facilitating role within countries, acting as the convening point for a national network of actors across the spectrum of those involved in relevant or potential socio-technical innovation systems (from users, through supply chains, to NGOs and policy-makers) and championing the development of socio-technical innovation systems around different technologies. Their core remit would be to link together national actors around a strategic, long-term, nationally defined vision that is cognisant of national policy goals and local realities. They would develop detailed knowledge of national capabilities and key areas where opportunities exist for rapid development and growth, and they would identify areas where international expertise and knowledge sharing are required. SEA-RIBs would provide strategic oversight, advising on how to target sustainable energy access (or broader climate technology) programmes and projects in a coordinated way that responds to identified priority areas for both rapid growth and long-term capability-building. As implementers of Proposal 2 below, such institutions would also work to ensure that all projects and programmes are used strategically to build and strengthen socio-technical innovation systems.

Careful attention will be needed from the outset to ensure that activities conform to the funding criteria of potential funders (e.g. donors, the development banks, GEF, the Green Climate Fund). This may require specific tailoring and packaging of different initiatives accordingly. The key added value of such funding being channelled through, or at least engaging with, SEA-RIBs is the opportunity to increase coordination and ensure every dollar spent leverages further benefits in building relevant aspects of national socio-technical innovation systems via a grounded understanding of the context-specific needs of individual countries and technologies. This would provide the most powerful and effective means of mainstreaming Socio-Technical Innovation System Building activities in individual countries, with myriad benefits in terms of driving transformations in sustainable energy access.

It is likely that SEA-RIBs would meet with the most success if they were situated in existing organisations in respective countries, preferably organisations with some level of existing knowledge of sustainable energy access and development. Such organisations would need a broad policy-level perspective on these issues as opposed to specialising, for example, in technology hardware and R&D, or finance. It should be evident from the analysis in this book that what is required is a perspective that goes well beyond the traditional narrow focus on technology and finance.

As outlined above, as with Proposal 2, it is easy to imagine the adoption of SEA-RIBs under international initiatives around sustainable energy access. If it were considered of value to connect with efforts around climate technologies under the UNFCCC, this could be achieved by framing sustainable energy access within the broader agenda around climate technology transfer and development. This could then entail using CRIBs (the climate-technology specific version of this proposal, described in Ockwell and Byrne 2015) as a means to strengthen the capabilities of the NDEs that currently are the only national level presence that the CTCN has. NDEs usually represent a small percentage of a civil servant's time – nothing like the level of dedicated institutional presence and resources required to engage in meaningful Socio-Technical Innovation System Building. The introduction of CRIBs could work well within the UNFCCC, as they could be Party-led, responding directly to requests for support from specific countries and delivered via a simple extension of the existing architecture of the UNFCCC's Technology Mechanism.

It is important to note that the concept of a centre-based approach to Socio-Technical Innovation System Building has synergies with, but differs in important ways from, existing centre-based ideas, both in the literature and in practice. A centre-based approach formed the central thrust of at least two proposals at the time of the critical UNFCCC negotiations in Copenhagen in 2009, in the form of a policy brief by Ockwell *et al.* (2009) and, most notably and more substantively, a paper by Sagar *et al.* (2009). Both called for the establishment of 'climate innovation centres' in developing countries, citing the successes of initiatives such as the CGIAR and the UK Carbon Trust (Sagar *et al.* 2009) and institutions such as Innovación Chile (see Ockwell *et al.* 2010b). Significantly, Sagar *et al.*'s paper led to infoDev's commissioning further analysis by Sagar (see Sagar and Bloomberg New Energy Finance 2010), which led to the establishment of infoDev's CICs. Importantly, however, the CIC approach differs in practice from the approach suggested by Sagar *et al.* (2009), which had as its central tenet the use of CICs to build national innovation systems.

Sagar *et al.*'s (2009, p. 280) proposal was for 'a network of regional "Climate Innovation Centres"' that would focus explicitly on building 'innovation ecosystems' around specific low carbon energy technologies (note: they referred to innovation ecosystems while citing the literature on national systems of innovation). This included a range of capacity-building activities – activities that go well beyond what eventually became the infoDev-led CICs. The CICs also differ from Sagar *et al.*'s proposal in that they are nationally situated, not regional. But, as discussed above, their activities are far more limited and focus very much on financing the activities of entrepreneurial SMEs, ignoring the activities of the multitude of other relevant actors who make up national innovation systems, or the kinds of activities that nurture them.

As emphasised in Ockwell and Byrne (2015), there are important differences between the CRIBs and SEA-RIBs proposed here and either Sagar *et al.*'s (2009) proposal or any of the existing international policy initiatives around sustainable energy or climate technologies. First, both CRIBs and SEA-RIBs are intended to operate at the national and sub-national level and reach out from here to the regional or international level. This responds to the emphasis in the literature reviewed in Chapter 2 and Chapter 3 on national and sub-national level interventions. For example, the actors identified in Chapter 4 and Chapter 5 who played a Socio-Technical Innovation System Building role in Kenya's solar PV sector were nationally situated actors, most having been present in the country and focussed on work on solar PV for decades. A similar knowledge of national circumstances and capacities has been observed in China's strategic use of the CDM to strengthen its own national innovation system (Watson *et al.* 2015). Indeed, it is difficult to find innovation system level analyses that focus at any level above the national level. Most examples cited in the literature on climate change and innovation are national-level interventions – the UK's Carbon Trust, Chile's Innovación

Chile. Only the CGIAR represents a regional-level initiative, although arguably much of its success was achieved by targeted interventions during sustained periods of national presence.

Second, CRIBs or SEA-RIBs are intended to focus on the broader socio-technical innovation systems as described in this book. This goes beyond the scope of Sagar *et al.*'s (2009) definition of an 'innovation ecosystem' or the conventional definitions of national innovation systems within the Innovation Studies literature. Rather, it focuses on the extended understanding developed in Chapter 2 and Chapter 3, which uses insights on innovation system building from the Innovation Studies literature and brings these to bear in the context of Socio-Technical Transitions theory. The latter attends, in particular, to the social practices of technology users and the existing socio-technical regimes, both of which shape the enabling and competitive environments within which sustainable technologies must survive and, hopefully, thrive.

Proposal 2: Using projects and programmes to build socio-technical innovation systems

Ideally, Proposal 2 would be pursued in tandem with Proposal 1 as part of the strategy of SEA-RIBs in boosting Socio-Technical Innovation System Building across countries. Proposal 2, can, however, viably pursued independently. Actors engaged in any sustainable energy access-related activities could adopt the approach described here as a means to ensure the maximum impact of their individual projects and programmes. The more they coordinate such efforts with other actors, the greater the likely impacts.

Proposal 2 essentially involves mainstreaming Socio-Technical Innovation System Building across all sustainable energy access projects and programmes, ensuring every opportunity is taken to use projects and programmes to achieve broader Socio-Technical Innovation System Building impacts. This requires mainstreaming a focus on building innovation systems across all projects and programmes and designing and implementing them as real-world experiments in which to better foster learning, and capability and system building. The specifics of how projects and programmes can be used as opportunities for Socio-Technical Innovation System Building, in line with the overarching policy goals articulated above, are outlined in more detail below.

From the evidence and analysis presented in this book, it is clear that there is a role for donors (and other funders, including inter-governmental organisations and NGOs) in such projects to provide adequate protection against the full force of market selection pressures. It is under these conditions that stakeholders can experiment to generate the learning needed for the sustained development, transfer and diffusion of sustainability-related technologies and practices, and to nurture the development of socio-technical innovation systems. But there are other aspects to the design of projects and programmes that appear to be important. First, we should be clear about what a project or programme is meant to achieve. Is it the demonstration of a ready-made solution for others to imitate, or is it experimentation to contribute to understanding of what solutions could work? Second, the motivation of project participants needs to be considered, as does, third, the scope of projects. And, finally, the way in which projects relate to each other can have powerful impacts, which also generates implications for the role of institutions at national and international levels. Each of the aspects related to projects, donors and other public funding bodies, as well as national and international institutions, is elaborated below. Included in these elaborations are non-exhaustive suggestions of how each aspect of projects might relate to the four goals recommended above, underlining the importance of the interrelatedness of the goals as we emphasised before: (1) build networks of diverse stakeholders; (2) foster and share learning; (3) promote the development of shared visions; and (4) support diverse experimentation.

Projects as experiments

Projects and programmes should be seen and used as experiments that are implemented in order primarily to learn, rather than aiming solely to achieve or demonstrate particular solutions. In other words, they could be recast as experiments to make this learning function clearer, in a similar sense to the way R&D activities are often characterised. As such, the measures of success of a project (or programme, experiment) need to be considered carefully. Quantitative indicators can be useful but they can become the sole focus of evaluation. A range of qualitative 'indicators' could help to identify more subtle but important impacts, such as the kinds of knowledge created from experimentation or the nature of relationships fostered in network-building. This could also help to reduce the tendency to assess projects and programmes in 'failure' versus 'success' terms, thereby encouraging the sharing of outcomes. In essence, this is about the need to redefine success as the generation of important lessons, rather than ready-made solutions.

In terms of the four goals recommended above, this aspect of projects most clearly relates to supporting diverse experimentation (goal 4). But the purpose of experimentation, as has been argued, is to create opportunities for learning, and so there is a direct link to the goal of fostering and sharing learning (goal 2). That is, the experiments themselves are the spaces in which learning is fostered. However, learning is only useful to broader innovation system building if it is shared. These lessons will, of course, be immediately available to project participants who, by working together, will form a network (at least for the duration of the project) and thereby contribute to network-building (goal 1). But, for wider and longer-term network and innovation system building, lessons need to be shared publicly. This will not only help to build networks of diverse stakeholders (by providing lessons of potential interest to actors external to projects themselves), but it can also promote the development of shared visions by grounding possible visions in real-word experience (goal 3).

Motivation of project participants

In order for projects and programmes to generate useful learning, the participants must be motivated to solve real problems. That is, the problems the project or experiment explores need to be relevant to those involved and so should be defined by them. The motivation will be further enhanced if the participants have material interests in the outcomes – if the learning will have value for them. There is a clear link here with the issue of risk. While mitigating risk is important, particularly for private sector actors, the elimination of risk could be de-motivating. So participants should be expected to invest some material resources in experiments, partly to demonstrate to others their commitment but also to ensure that they have a stake in the outcomes.

This aspect of projects highlights the need for them to be attractive to potential participants and so, considering the goal of building diverse stakeholder networks (goal 1), reinforces the point above that problems should be defined by potential participants. Moreover, this self-definition of problems will raise the chances that projects will be both relevant to diverse stakeholders and create opportunities for learning from a diversity of individual perspectives and particular contexts. Clearly, there are links to fostering and sharing learning (goal 2). But responding to participant motivations for project involvement is also more likely to mean deeper commitment to projects and attempts to develop shared visions (goal 3). And, if attempts to attract a wide variety of participants are successful, then there will be more opportunities to conduct a diversity of experiments, thereby linking with goal 4.

The scope of projects

It is clear that learning is facilitated by deep interactions among a broad range of actors who can bring their problem-solving efforts to bear on the many dimensions of development pathways as they unfold in different contexts. This suggests that there needs to be experimentation on many of these dimensions simultaneously (and links with our notion of systemic intervention). However, it would be extremely difficult for a small number of actors to achieve this. To overcome this difficulty, either complex projects involving a wide range of stakeholders could be implemented or many simpler projects could be implemented programmatically, each one operating on a selection of the dimensions of a development pathway. Each approach will have its advantages and disadvantages. The point is to generate learning across the multiple dimensions of a pathway so that sustainability-related technologies and practices can emerge in a co-evolutionary process. The assumption here is that co-evolutionary learning will tend to produce mutually reinforcing technologies and practices that operate in sympathy with their context, thereby increasing the chances of widespread adoption of those technologies and practices – and their sustainability.

Another important point here relates to continuity of efforts. Here, programmes may have the potential to deliver innovation system building in ways that individual projects may not. Funders often want to see results within a few years. Although funders should monitor progress and stop activities when they are clearly not functioning, really making headway on an innovation system might take much longer than a project period – although the potential contribution of individual projects should not be underestimated. Nevertheless, unless within a programmatic context with a timespan of, say, ten to fifteen years, or within the context of a more coordinated national approach to commissioning projects (as would be achieved via the creation of the SEA-RIBs advocated in Proposal 1 above), projects run the risk of being one-off efforts with limited structural contributions. A related point is a trusting relationship between different actors. In societies where contracts do not play a huge role but relations make the difference, having the same person run the same programme (or SEA-RIB) for longer can be a key success factor.

In terms of the recommended goals, projects (or programmes) with a wide scope – as indicated by the range of development dimensions along which a project or programme is operating – are more likely to result in a diversity of learning opportunities and lessons generated. Most clearly, this links with the goal of fostering and sharing learning (goal 2). And, of course, this links clearly with the recommendation to support diverse experimentation (goal 4). But projects with wide scope are also likely to need to engage with a wide range of actors, and so they increase the opportunities to build networks of diverse stakeholders (goal 1). If there is support for projects and programmes over the longer term – as per the point above about continuity of efforts – then there is also more chance that such networks will develop strong relationships (also contributing to goal 1). The combination of learning from diverse experimentation and the continuity of network building should also help actors to develop shared – and grounded – visions (goal 3).

Interactions with other projects

Following on from the previous recommendation, even complex projects or programmes of projects could be constrained in their learning, particularly if the funding is from a narrow range of sources. Moreover, if they are under the same management, they will be dependent on the particular abilities of that management. As the case study explored in this book demonstrates, projects or programmes implemented from different perspectives, if encouraged to interact meaningfully over the long term, can generate learning that helps to achieve significant results. This requires some degree of coordination, of course, but not necessarily management. That is, the individual projects and programmes need to be able to communicate directly with each other as well as via a central actor. It is here that value could be added by making SEA-RIBs internationally networked, e.g. through the UNFCCC's Climate Technology Centre and Network (especially in the CRIBs version of our proposal).

Encouraging interaction across projects clearly links with the recommendation to foster and share learning (goal 2) but there are also links to the other goals. Interactions will help to further build networks of diverse stakeholders (goal 1) by creating opportunities for various stakeholders to meet and share their knowledge. But interactions of this kind can also create spaces in which stakeholders discuss, debate and develop shared visions (goal 3). And awareness and understanding of other projects mean the possibility to ensure that any new projects or programmes do not replicate unnecessarily experiments already conducted, thereby contributing to the goal of supporting diverse experimentation (goal 4).

Role of donors and other public funding

Many private sector actors, particularly small players in developing countries, cannot risk much of their capital to undertake experiments. However, there might be significant benefits if they were able to do this, for them and for wider society. Therefore, a substantial share of the risk inherent in experimentation could be borne by donors, who can justify their support in terms of these potential social benefits. Other sources of public funding, including the Green Climate Fund and the regional development banks, could serve a similar purpose – although it is important to ensure that funding sources are also accessible to smaller actors who might not have the capacities to engage with large, multilateral funding streams (suggesting a role for donors and NGOs in bridging or plugging this gap). The involvement of public funding also has the additional significant benefit of making learning from projects publicly available, thus contributing to wider learning and long-term capability building.

Another aspect of the risk issue is the stability and long-term provision of support, as noted above in regard to the continuity of efforts. If the support is unstable, intermittent or short term, then it is more likely to increase risk than mitigate it. This is not to argue that support should be unconditional. There needs to be a way to maintain motivation in individual projects but the thematic, or overarching, support can be maintained so that there is confidence among stakeholders that it is worth their investing effort in particular experiments.

Linking with the recommended goals, we can see that the risk-bearing nature of public funding will more likely foster learning (goal 2) because of the space it creates in which to experiment (goal 4). And public funding means a greater likelihood to share learning because of the demand to make available publicly-funded research (goal 2). But the public availability of lessons can also help in building wider networks of stakeholders (goal 1). And wider availability of learning can help in public discussions and debates about shared development visions (goal 3).

Role of institutions

In order to achieve all of the above in a way that maximises the potential impacts in terms of Socio-Technical Innovation System Building, appropriate institutional structures are necessary. It is this that drives the rationale for the creation of SEA-RIBs under Proposal 1 above. In the absence of such central, nationally-based institutions, organisations seeking to operationalise Proposal 2 would need to look at ways of mainstreaming such Socio-Technical Innovation System Building through their own approaches to developing, supporting, monitoring and evaluating projects and programmes.

Finally, with regard to the recommended policy goals above, institutions of the kind discussed can provide formal channels and mechanisms for coordination and linking. So, institutions can link to other institutions in formal arrangements, whether they are sub-national, national or international. This directly helps to achieve network building (goal 1). It also helps to coordinate the sharing of lessons from projects (goal 2) and, indeed, can be useful for the coordination of projects and programmes themselves, such that there is a continuing diversity of experimentation (goal 4). And, in exploiting formal links and stakeholder networks, institutions can organise more structured forums in which to develop shared visions (goal 3).

Conclusion

Building on the theoretical developments in Chapter 2 and Chapter 3, and the empirical analysis in Chapter 4 and Chapter 5, this chapter has demonstrated a number of things in relation to policy and practice. First, clear limitations are evident in the extent to which Hardware Financing interventions, such as PVMTI, and Private Sector Entrepreneurship interventions, such as the CICs, have achieved or, in the case of the CICs, are likely to achieve transformative impacts, in the future. Second, there is clear potential for interventions such as the CICs, as well as the other international policy initiatives described above, to broaden their aims and activities in order to more effectively engage in interventions that are likely to have wider impacts through contributions to Socio-Technical Innovation System Building. Finally, and critically, the analysis above has articulated a number of key goals for policy interventions aimed at building socio-technical innovation systems, together with an elaboration of more specific interventions that would support the achievement of such goals. As is clear from the discussion above, there is no reason why such interventions need be limited to a focus on sustainable energy technologies – they could (if desired) be levelled more broadly at climate technologies or other technologies of relevance to different aspects of sustainable development.

While we stand by the policy proposals articulated above as positive ways forward in relation to sustainable energy access and broader attempts to achieve low carbon or climate-compatible development and 'green growth', it is important to flag at this point some critical caveats. As is dealt with more centrally in Chapter 7, proposals such as the two detailed above raise important issues relating to the governance of the kinds of transformations in access to sustainable energy technologies that initiatives such as SE4All and discourses around ideas like green growth imply. Implicit in both of the policy proposals above is some sense of organisations or individuals who have the agency, closely related to financial and political resource availability, to drive the establishment of the kinds of institutions that we advocate (whether CRIBs, SEA-RIBs or some other version). This also applies to any actor who would change their practices in order to implement activities around Proposal 2, implementing strategic changes to their projects and programmes to maximise potential for Socio-Technical Innovation System Building. We might usefully refer to such actors as *Socio-Technical Innovation System Builders*. It was clear from the empirical analysis in Chapter 4 and Chapter 5 that key actors did indeed play a central role in driving the development of the Kenyan solar PV market, building socio-technical capabilities around solar PV, lobbying government and donors, connecting different actors. But, when thinking about applying this idea to other technologies, practices, services, and so on, who should these actors be? How can we ensure that when Socio-Technical Innovation System Building is pursued, it does not play out in ways that serve the interests of powerful elites as opposed to the poor people who, it is assumed, will benefit from sustainable energy access? Who gains? Who loses? Beyond the pro-poor focus at a national scale, how will these dynamics play out globally? How can we be sure a similar story as was observed with the CDM is not observed with sustainable energy access, with benefits flowing to emerging economies like China and India as opposed to low-income countries like Kenya? Will the kinds of nationally situated, potentially more engaged and participatory institutions that we envisage in our proposals really emerge in practice? If they do, can they overcome the entrenched political economies within and across nations to truly deliver pro-poor transformations in technology development, transfer and access? These issues concerning the politics of any transformation in sustainable energy access, and the questions of governance raised by the analysis in this book and the policy proposals above, are revisited in Chapter 7.

Notes

1 SolarNet was a network for renewable energy promotion in the region and published a widely read newsletter a few times per year. It was formally closed down in 2010 (Kilonzo 2013, interview).
2 Ten per cent of PVMTI money was already available for grants for exactly the kinds of activities the stakeholders wanted funded (IFC 1998). It is unclear why it took so long for the money to be made available in-country. But additional grant money was made available after the grant component was increased to 20 per cent (IFC 2007b; Ngigi 2008, interview).
3 A wide range of materials is available on the Lighting Africa website: www.lightingafrica.org/
4 See www.infodev.org/articles/climate-technology-read-more-about
5 Taken from www.kenyacic.org/?q=node/59
6 Taken from www.infodev.org/articles/climate-technology-read-more-about
7 Taken from www.kenyacic.org
8 Taken from www.kenyacic.org/?q=node/17
9 See www.unep.org/climatechange/ctcn/Home/tabid/131937/Default.aspx
10 For a full list, see http://unfccc.int/ttclear/templates/render_cms_page?s=TEM_ndes
11 See http://unfccc.int/resource/docs/2013/sbi/eng/05.pdf and http://unfccc.int/resource/docs/2014/sbi/eng/inf03.pdf (Appendix 1).
12 See www.adb.org/sites/default/files/pub/2012/pilot-asia-pacific-climate-technology-flyer.pdf
13 See www.afdb.org/en/news-and-events/article/afdb-creates-african-pilot-climate-technology-and-finance-centre-with-gef-support-13344/

14 See www.iadb.org/en/projects/project-description-title,1303.html?id=RG-T2384
15 These proposals are drawn from Ockwell and Byrne (2015) in which they refer to Climate-Relevant Innovation system Builders (CRIBs). Here, we have adapted the acronym to reflect the fact that our primary focus in this book is on sustainable energy access.

7

CONCLUSION

Towards Socio-Technical Innovation System Building

We began this book by articulating the enormity of the energy access problem in Sub-Saharan Africa and calling for new scholarly thinking that matches the ambitions of contemporary policy commitments like SE4All. In this concluding chapter, we take stock of the contributions of the book from both an academic and a policy and practice perspective. Our aim is both to re-emphasise the arguments for which the book has provided empirical evidence and to suggest some of the questions that we consider still remain to be answered. We also take some time to engage with a potential critique of the book's analysis, related to its focus on off-grid solar PV in Kenya. Undoubtedly, as the agenda for future research we articulate further below emphasises, there is an urgent need for comparative research that explores the nature of our findings in relation to experiences of transformative – and non-transformative – change across different contexts. Our own research, however, and that of others in some of these different contexts, suggests that the Socio-Technical Innovation System Building perspective articulated in this book has considerable traction across different contexts. This supports the likely transferability of the insights in this book and the catalytic policy mechanisms that we have articulated. It demonstrates why the case of off-grid solar PV in Kenya cannot easily be dismissed as a 'unique and isolated exception' (Fairhead and Leach 1996, p. 280).

Notwithstanding the potentially broad applicability of the book's findings, this concluding chapter is as much an agenda for further research as it is a summary of the book's main arguments. Of course, the book raises as many questions as it answers. Our hope, however, is that by providing a more systemic perspective from which to understand the adoption of pro-poor sustainable energy (and other) technologies, the questions raised move us forward towards a policy and research agenda better equipped to transform pro-poor energy access – an agenda that takes us well beyond the literature's traditional two-dimensional focus on technology/engineering and finance/economics.

We begin this chapter by summarising the book's key empirical, theoretical and policy and practice contributions. We then move on to consider the issue of how transferable these findings are beyond the case of off-grid solar PV in Kenya. The book then concludes by articulating an agenda for future research.

Socio-Technical Innovation System Building and pro-poor green transformations

As noted in Chapter 1, 'transformation' is an over-used word, particularly in contemporary academia in relation to sustainable development. But many international policy commitments imply nothing short of a transformation in the availability and use of sustainable technologies, including the UN's target of providing sustainable access to energy for an additional 1.1 billion people over the next 15 years, as well as multiple sustainable development goals (SDGs), international climate change commitments and continental-scale strategies framed around Green Growth and ideas of Low Carbon or Climate Compatible Development. As we have seen in the decades of empirical work in the field of Innovation Studies, and recent advances in thinking about Socio-Technical Transitions, such transformations require us to go far beyond simplistic notions of financing the adoption of technological hardware. Neither will such transformations likely be driven by increasing the availability of venture capital or business incubation services for individual green enterprises. Any deliberate intervention to try to drive transformations in sustainable technology availability and adoption needs to be both systemic in nature and able to engage in co-evolutionary interactions with the social practices that sustainable technologies might facilitate, and to tackle the interlocking political, economic and socio-cultural path dependency of incumbent socio-technical regimes.

By focussing our analysis at the level of a national socio-technical innovation system (i.e. the market for off-grid solar PV in Kenya), we have been able to go beyond the usual project-by-project (or programme-by-programme, policy-by-policy) 'barriers' analyses that characterise much of the existing literature on energy access in Sub-Saharan Africa. Arguably, this project-by-project empirical focus is also a characteristic of some of the more insightful and socio-culturally attuned recent contributions to the literature (e.g. Winther 2008; Sovacool *et al.* 2011; Ulsrud *et al.* 2011; Ahlborg and Sjöstedt 2015; Ulsrud *et al.* 2015). While a project-by-project approach may be appropriate for drilling down into detail in relation to specific issues (e.g. focussing on gender-based considerations or specific changes in socio-cultural practices enabled by electricity access, as per the detailed ethnographic work of Winther 2008), it is limited in the extent to which it can comment at the level of what might be considered transformational.

In contrast, by focussing at the national level, and on what is widely described as one of the biggest successes in the availability and adoption of off-grid renewable energy technologies in Africa, this book has been able to offer a more systemic perspective on how pro-poor green transformations might be achieved. This goes well beyond the catch-all term 'enabling environment' that is seen repeatedly in the literature, but is rarely, if ever, theorised in any detail. We have, as far as possible in this book, been as specific as we can about the nature of the 'environment' in which energy access, and socio-technical innovation and change more broadly, can be nurtured. Through detailed historical analysis of the Kenyan off-grid PV market, the analysis in this book has demonstrated how socio-technical innovation systems might be understood to represent the kind of enabling environment within which sustainable technologies become more widely available and supplied in ways that are better attuned to the social practices and needs of potential users. The term 'socio-technical innovation system' might be a bit of a mouthful compared to 'enabling environment' (suggestions for alternative terms are welcome!), but the former goes a lot further than the latter in specifying the conditions under which changes in technology availability and adoption occur, particularly when the technologies and users concerned are marginalised (i.e. sustainable vs. conventional technologies and poor vs. wealthy consumers) and unlikely to be served in the absence of deliberate policy intervention.

As is the case with Lighting Africa, we have also shown how policy approaches might be specifically designed to take a Socio-Technical Innovation System-based approach to driving transformations in the adoption of sustainable technologies among poor people. The analysis in Chapter 6 showed how it is possible to use the component parts of our theoretical framework to articulate specific policy goals and design interventions with the potential to nurture pro-poor socio-technical innovation system building around particular technologies or social priorities. As with our theoretical approach more broadly, it is possible to design such interventions to focus on socio-technical innovation beyond just sustainable energy technologies. This emphasises the relevance of Socio-Technical Innovation System Building across the suite of contemporary policy areas that require innovation in both technological hardware and socio-cultural practices around technology use – a relationship which the Socio-Technical Transitions literature (reviewed in Chapter 3) helpfully demonstrates to be co-evolutionary, and one that the Innovation Studies literature (reviewed in Chapter 2) also demonstrates to be systemic. Hence, the Climate Relevant Innovation-system Builder (CRIB) and Sustainable Energy Access-Relevant Innovation System Builder (SEA-RIB) policy approaches described in Chapter 6 represent a model for potential intervention that can deliver across a broad spectrum of sustainability concerns where technology and innovation have a central role to play within normative commitments to delivering social justice.

This latter concern with social justice is fundamental to the book's focus on explicitly *pro-poor* green transformations. As we have demonstrated, Private Sector Entrepreneurship or Hardware Financing policy framings ignore fundamental aspects of how socio-technical change occurs. As such, they are not fit for purpose as stand-alone approaches to attempting to deliberately intervene deliberately through policy and practice to drive green transformations that will deliver against the needs of the poor. This works at a human development level, where poverty alleviation among individual women and men, and households and communities, are of concern. It also works at more aggregate economic development levels. There is little to suggest that a hardware financing-based mechanism like the Clean Development Mechanism (CDM), or a private sector entrepreneurship-focussed approach like the Climate Innovation Centres (CICs), will do much to transform the economies of low-income countries like Kenya. Both the theoretical discussion and empirical analysis in this book clearly demonstrate that such transformations evolve through interventions at a systemic level, facilitating the development of appropriate capabilities around targeted technologies and specific social practices. Our argument throughout has been that a Socio-Technical Innovation System Building-based approach is more attuned to the way in which innovation and technological change can happen so as to underpin both human and economic development. By using this systemic understanding of innovation, technological change and development to expand thinking around socio-technical transitions – with its sharper focus on social practice and socio-technical regimes – research, policy and practice can become far better equipped to interpret and intervene in ways that can facilitate pro-poor green transformations.

In this way we see how framings matter and how the politics of who frames problems have material consequences for who gains and who loses (Leach *et al.* 2010a). As we elaborate below, however, there is still important work to be done in analysing the politics of pro-poor green transformations. And all of this has significant implications in terms of questions of governance. Before discussing this, however, we first deal with an important question regarding the transferability of the lessons learned from this book beyond the context of off-grid PV in Kenya

Beyond Kenyan off-grid solar PV

This book has focussed exclusively on an in-depth historical account of the development of the off-grid solar PV market in Kenya, which is claimed to be the largest per capita market for off-grid PV globally. Bearing in mind the significance of this market, the detailed account we have given of its evolution – focussed on the minutiae of efforts over many years around a single technology – is arguably one of the key strengths of this book. The widespread understanding of the Kenyan PV market phenomenon frames it as a free market, or private sector-led, success story. Our account challenges this understanding and suggests there are serious policy implications that flow from this alternative view. However, our approach also raises an important question – and potential critique – relating to the extent to which the theoretical insights and lessons for policy and practice are transferable across different countries and different technologies.

Certainly, learning lessons from Kenya and applying them in different contexts ought not to be dismissed outright. Such 'learning from abroad' offers important opportunities for countries to learn at considerably less political cost to themselves than repeating the mistakes of others. These lesson-drawing approaches to public policy-making form the fundamental tenet of a school of public policy analysis typified by authors such as Richard Rose. As Rose (2004, pp. 1–2) puts it:

> In a world in which people, money, and ideas increasingly move across national boundaries, ... national policymakers [need] to abandon the belief

that the only wheel worth using is one that is invented at home. Policymakers can learn how improvements might be made by looking elsewhere. After all, that is how Japan turned from being a country that imported automobiles to being the world's biggest exporter of cars.

Nevertheless, it is obvious that both the theoretical assertions and lessons for policy and practice we have offered in this book would be strengthened through comparative analysis. Such comparisons could be across different countries, technologies, levels or 'stages' of technological development, scales of technological application, urban and rural applications, and the different social practices that different technologies might facilitate. All these considerations have important implications for the ways in which different technologies might interact within the context of specific socio-cultural and political and economic landscapes. We should therefore acknowledge the importance of such context-specificities (Ockwell and Mallett 2012) and offer some concrete empirical examples that support the Socio-Technical Innovation System Building perspective we have developed in this book. In the following paragraphs, therefore, we present a range of insights from both our own research and the research of others, across a range of different developing country contexts and different technologies. All of this gives important clues as to the potential traction that Socio-Technical Innovation System Building offers, both theoretically and for interventions through policy and practice. Moreover, the catalytic nature of the policy approaches we proposed in Chapter 6 is designed specifically to be able to respond to context-specific considerations across different country, technology, socio-cultural and political landscapes.

Let us begin by considering China, whose astounding success in relation to various climate technologies (e.g. wind and solar) is well known and often cited. China is often dismissed as a special case, a country well ahead of most other developing countries in terms of its levels and rate of economic growth, and the technological capabilities it possesses. This, together with its sheer size, means it is unique – too different from other countries to draw comparable lessons. However, we would argue that, while rapidly increasing in sophistication, China's current capabilities across many technologies are not as advanced as is often assumed (e.g. see Breznitz and Murphree 2011; Watson et al. 2015). A closer look at the history of how China has developed capabilities across different climate technologies reveals a story that is of significant relevance to other developing countries in considering strategies for galvanising more rapid rates of technology development, transfer and diffusion (together with the opportunities for economic growth that this can bring). The story is one of strategic, systemic and largely incremental development of indigenous technological capabilities via a careful approach to nurturing China's innovation system. As Stua (2013) demonstrates, even in its engagement with international policy mechanisms like the CDM, China has pursued a careful and deliberate strategy of using these mechanisms to bolster domestic efforts to build technological capabilities and strengthen its national innovation system.

Watson *et al.* (2015) provide an insight into China's innovation system building using various low carbon technologies through empirical analysis across three sectors: (1) energy efficiency in the cement industry; (2) electric vehicles; and (3) efficient coal-fired power generation. In each of these sectors, similar patterns are observed, yielding several insights to design a policy that aims to boost low carbon technology transfer, development and diffusion in other developing country contexts, for example:

1. Rather than technological capability-building being driven 'downwards' by efforts in complex R&D-led interventions, it tended, in the cases examined, to have been driven 'upwards' from technological learning at the simpler market-oriented level. In cases where policy focussed on R&D-driven interventions to build capabilities (e.g. in electric and hybrid electric vehicles), there had been less success in the development of Chinese firms' technological capabilities. This supports Bell's (1997, p. 75) assertion that 'dynamic technological capabilities are cumulatively built "upwards" from simpler to more complex design, engineering and managerial competences, not "downwards" from R&D'. The focus at the market level also provides important clues as to the potential significance that a socio-technical perspective might yield to such analysis (Watson *et al.* adopted a solely Innovation Studies-based theoretical approach).

2. Where Chinese firms were observed to have made advances, this had mostly been achieved via relationships with foreign firms to facilitate knowledge flows and learning. In several cases, this was achieved through relationships with 'second-tier' companies rather than those at the technological frontier. Initial advances typically developed production capabilities in Chinese firms. These led to learning and incremental change, and the later development of more advanced innovation capabilities.

3. Chinese domestic policy played a key role in incentivising firms' engagement with low carbon technologies. A combination of market support, regulations and R&D support was crucial, creating a more systemic approach to low carbon technology transfer, development and diffusion, and the development of increasingly advanced technological capabilities.

4. The Chinese government took a strategic and systemic approach to leveraging opportunities for funding and capability-building through international climate change policy, using the CDM in particular as a means to sharpen domestic policy efforts in order to strengthen its national innovation system in relation to the sectors analysed. This innovation system building approach was clearly more effective in building indigenous capabilities in China than an international policy mechanism (the CDM) that supports hardware transfer on a project-by-project basis can be in isolation. In other words, the Chinese government built institutional capacity focussed on leveraging the CDM to build China's innovation system.

These insights are significant. They demonstrate how a strategic policy-driven approach, which focusses at a systemic level, building national innovation systems through bottom-up and often incremental processes of technological capability-building, can lead to much more rapid and sustained processes of sustainable technology transfer, development and diffusion – with myriad accompanying economic benefits. In many ways, the case of China can be seen as a master class in innovation system building around sustainable technologies. This kind of system building-focussed approach is an excellent example of the kinds of catalytic interventions that the CRIBs and SEA-RIBs proposed in Chapter 6 could facilitate. The key caveat to the China story above is that our intention is for CRIBs and SEA-RIBs to be governed in ways that are explicitly pro-poor, which is different from the industrially focussed strategy observed in China (see the discussion of governance in the final section of this chapter below).

Another insightful case study is the market for SHSs in Kenya's neighbour, Tanzania (see Byrne 2011). In 2004, the IFC launched a GEF-funded programme (working through the UN Development Programme, UNDP) which, like PVMTI in Kenya, aimed to increase adoption of SHSs in Tanzania. In contrast to the Kenyan experience under PVMTI, the Tanzanian case study provides an insight into the potential success of schemes that focus on capacity building from the outset, as opposed to hardware financing. Based in Mwanza Region, near Lake Victoria, the project made USD 2.5 million available and chose to focus at the level of government energy policy with an aim of building capacity and creating markets around SHSs. There were five main interventions, namely: (1) policy influence (technical standards, lower duties and taxes); (2) private sector capacity building (technical and sales); (3) raising awareness (demonstrations, advertising); (4) enhancing affordability (using micro-finance); and (5) replication in nearby regions. The project met with a high degree of success in building innovation capabilities around SHSs in the region, as well as having broader policy influence, thus strengthening a key element of effective innovation systems. It facilitated a standards-setting process (in collaboration with the standards-setting process in Kenya under the later stages of PVMTI). It also persuaded government to reduce taxes and duties on SHS equipment. Other innovation capability-building successes, which speak directly to various concerns emphasised by a Socio-Technical Transitions perspective, included increased awareness of SHSs among the local population, thus enhancing the potential for adoption among technology users. In a matter of a few years, significant expansion in the market for SHSs in Mwanza Region was observed, contributing to rapid national SHS market growth and at the same time supported by project efforts in other parts of Tanzania. By 2008, the annual market for solar modules in Tanzania was estimated to be worth USD 2 million, with around 10,000–14,000 modules having been sold between 2006 and 2007.

Overall, the case of Tanzania, together with the later emphasis on capacity building activities under PVMTI in Kenya (described in Chapter 6) provide a good illustration of how a focus on innovation capability-building, and related innovation systems, can be extremely fruitful in fostering sustainable technology transfer and diffusion. It is also important to consider the longer-term benefits that will have accrued from these capability-building efforts beyond the lifetime of the projects.

For example, once the Kenyan PVMTI focussed efforts towards capacity building, confidence in the market for SHSs was observed to increase, countering a previous negative perception of the technology relating to poor quality components, a lack of skills and a lack of independent information about SHSs and supporting capacity.

One area where the Tanzanian project was not successful was in its experiments with micro-finance to enhance affordability of SHSs. This has been reported by some observers as being due to difficulties in securing high-level management support from the banks for SHS loan products, and the high risks associated with lending to dispersed rural customers (Byrne 2011). Recent analysis by the authors with Paula Rolffs, however, demonstrates how the failure of such traditional financing approaches can be better understood as being due to their not being properly attuned to the existing practices of poor people in paying for and consuming energy. Rolffs et al. (2015) provide a powerful insight into the relevance of a socio-technical perspective to understanding the access to and adoption of sustainable energy technologies among poor people (the paper operationalises a strategic niche management approach to its analysis). Focussing their analysis on the emergence of new mobile-enabled pay-as-you-go (PAYG) financing models for solar PV in Kenya, the paper demonstrates how the design of these PAYG approaches (facilitated by an interesting fusion between the widespread adoption of mobile banking across Kenya and the availability of cheap Chinese solar PV technology) has been developed through close attention to the ways in which people pay for and consume conventional energy sources, such as kerosene. The new PAYG finance models allow people to pay for solar PV via payment schedules that match the frequency and cost of existing energy payments (often being cheaper than kerosene payments). This attention to socio-cultural practice renders PAYG finance models more likely to succeed than traditional micro-finance models, which have largely failed across Kenya and Tanzania. Such analysis serves to further reinforce the relevance of a socio-technical perspective on understanding ways in which pro-poor access to sustainable technologies might be transformed. In this case, even though the paper's focus is on one of the two dimensions that dominates mainstream analysis in the energy access literature (i.e. finance), the use of a socio-technical approach yields fundamental insights as to the reasons for the relative success of new PAYG finance models – insights that would be impossible without such attention to socio-cultural considerations.

There is now a small number of examples within the energy access literature that operationalize socio-technical approaches both to analyse successful adoption and design practical interventions – what Gollwitzer et al. (2015) refer to as the 'socio-cultural turn' in energy access research. Ulsrud et al. (2011), for example, analyse the implementation of solar mini-grids in the Sundarban Islands in India. Their analysis demonstrates the co-evolutionary nature of the adoption of these technologies with the social practices of the people whose lives intersect with them, and the significance of this to designing effective interventions in practice. These insights are developed through later action research by Ulsrud et al. (2015), where a socio-technical perspective is operationalised in the design and implementation of village-level solar mini-grids in Kenya. In this case, an interdisciplinary team of engineers and social scientists, using a socio-technical approach to design and follow-up activities, was observed to increase the economic sustainability and technical functioning of the mini-grid. Again, the attention to social practice and socio-cultural context is demonstrated to be fundamental to the success of the intervention. Similarly, but at an analytic level, Ahlborg and Sjöstedt (2015) demonstrate the utility of a socio-technical perspective in understanding the elements that led to the successful implementation (including strong community buy-in and ownership of the management) of a small hydropower system in Tanzania. The work of Winther (2008), while based on a social anthropological approach, is also revealing in demonstrating the co-evolutionary nature of social practices and sustainable energy access. An emphasis on the centrality of socio-cultural considerations in determining the success of energy access projects and programmes is also prevalent across all the case studies (which cover multiple countries and technologies) considered by Sovacool and Drupady (2012). In earlier work on SHS adoption in Papua New Guinea, Sovacool and colleagues (Sovacool et al. 2011) also argue that separating out technical and economic barriers to SHS adoption from social and political barriers eludes their interrelated nature. They argue in favour of what they call a 'socio-technical systems'-based perspective, although they develop this from separate roots than we have developed in this book. At a more industrial level, Hansen and Nygaard (2014) also use a strategic niche management perspective to explain the emergence of a palm oil biomass waste-to-energy niche in Malaysia.

Similar to this nascent literature that operationalises a socio-technical perspective in relation to sustainable energy access, there is an emerging literature – beyond that described further above – that develops the application of Innovation Studies-based perspectives, including innovation systems perspectives, to understand the development of markets and firm-level technological capabilities around various sustainable technologies in developing countries. For example, Hansen et al. (2015) operationalise a technology innovation system perspective to analyse regional policy developments in the markets for solar PV across Kenya, Tanzania and Uganda. Hansen and Ockwell (2014) also use an Innovation Studies perspective, focussed on technological capability-development via systemic interactions between domestic and overseas firms, to explain the development of the biomass power equipment industry in Malaysia. There are also examples where such perspectives have focussed on explicitly pro-poor applications, such as in the emerging literature on inclusive innovation (e.g. IDRC 2011; Leach et al. 2012; Foster and Heeks 2013; Fressoli et al. 2014).

Furthermore, the clues as to the transferability of the insights developed through this book are not limited to sustainable energy technologies. This is clearly illustrated by Moussa's (2002) analysis of the potential for technology transfer in the agricultural sector in Sub-Saharan Africa. Moussa highlights a wide range of biotechnologies that might improve food production in Africa without increasing land and water demands from the sector. Moussa's conclusions, however, are not simply that funding is required to finance investment in these technologies; rather, a range of capability-building initiatives are emphasised which would be necessary to facilitate technology transfer and adoption. These include training, improving linkages between markets, storage and distribution systems, rural micro-finance availability and improved networking between research institutions, rural infrastructure providers and the private sector. While Moussa's report is not explicitly interested in the issue of green transformations (rather its remit is simply to increase agricultural productivity in Sub-Saharan Africa), the recommendations it makes relate directly to key components of effective innovation capabilities and have clear resonance with the ideas of Socio-Technical Innovation System Building developed in this book.

There are, then, plenty of reasons to believe that the various insights and theoretical and policy perspectives developed in this book through detailed analysis of the Kenyan solar PV market have value beyond this specific country and technological context. This might be via policy learning to inform interventions and approaches in other country contexts. Or it might be via the transferability of Socio-Technical and Innovation Systems perspectives across multiple other contexts. Our firm assertion is that the Socio-Technical Innovation Systems perspective developed in this book has significant relevance to analysis and action in seeking to drive pro-poor green transformations, both in relation to sustainable energy access and many other areas where access to sustainable technologies is of central relevance (e.g. in relation to international climate policy and across the whole suite of the SDGs). Nevertheless, there is still significant work to be done, both in further developing the conceptual perspective articulated in this book and in ensuring that the kinds of policy interventions described in Chapter 6 succeed in achieving pro-poor outcomes.

A research agenda: governance, democracy and pro-poor green transformations

Throughout this book we have argued for the relevance of a socio-technical innovation system-based approach to transforming access to sustainable energy. We have also articulated concrete policy aims and policy approaches for building socio-technical innovation systems around sustainable energy (or other) technologies. While we stand by the potential of this approach to deliver widespread sustained changes in poor people's access to sustainable technologies, there are, nevertheless, several key issues that remain to be addressed and that are under-developed, both theoretically and practically. Before concluding, we therefore focus on articulating the key areas which we consider to warrant further attention both in research and in policy and practice.

We have already emphasised above the fact that the analysis in this book could be strengthened further by conducting comparative analysis across different countries, technologies, energy services, social practices, scales, urban-rural contexts and so on. It would also be useful to do comparative analysis with examples where attempts at transformation have failed, for example, using a socio-technical innovation systems perspective to compare experiences of PVMTI in Kenya with those in India. The differential success of Lighting Africa in Kenya compared to Ghana

(Urama 2014) also provides fertile ground for testing whether the relatively well-developed Kenyan PV socio-technical innovation system, compared with the less-developed Ghanaian system, is an important explanatory factor or not. As already emphasised, however, the lack of such comparative analysis by no means undermines the book's assertions (see discussion on the transferability of the book's findings above).

By far the most critical issue that remains to be addressed relates to the *governance* of Socio-Technical Innovation System Building. As emphasised elsewhere in the book, there is an implicit assumption in policy ambitions such as SE4All that actors (be they individuals or organisations and governments) can intervene in order to transform poor people's access to sustainable energy. The policy approaches articulated in Chapter 6 are designed as ways through which such actors can catalyse the building of socio-technical innovation systems. But implicit in all this is the idea that such actors will act in ways that are in the interests of poor people and poor countries. This, however, masks many critical and essentially political questions regarding who these 'socio-technical innovation system builders' are or should be, how they can understand the needs and aspirations of poor people, how such system building can be governed in ways that benefit poor people rather than maintain or worsen uneven power distributions and so on. What is there to guarantee that CRIBs or SEA-RIBs will be governed in ways that serve poor people? How can we be sure that they will not be pressured into serving national political economic interests, favouring domestic technologies and businesses over imported ones, or favouring industrial interests over the interests of poor people (see Newell *et al.* 2014 for an insightful analysis of the political economy of renewable energy in Kenya)? If sustainable energy technologies remain heavily import-oriented, who is gaining and who is losing on an international scale? How do such processes contribute to economic development within poorer economies? In other words, how can we guarantee that the potential for the approaches developed in this book are, in practice, implemented in ways that catalyse transformations that are truly pro-poor?

Clearly, the question of who gains and who loses is far from given, even where the kinds of catalytic, nationally focussed and potentially context-sensitive kinds of policy approaches proposed in Chapter 6 are adopted. Such questions of who does the building of socio-technical innovation systems also point to an under-theorised aspect of the perspective we have developed in this book. Work needs to be done, based on the kind of in-depth historical analysis presented in this book, to articulate who the socio-technical innovation system builders are and the kinds of (political) work they do to build such systems. Clues seem to exist within both the Innovation Studies and Socio-Technical Transitions literatures. For example, some papers make reference to relevant ideas such as actors who serve to connect together different parts of an innovation system, referred to as 'systemic intermediaries' (van Lente *et al.* 2003) or those whose actions play a role in driving cumulative causation (often borrowing from Political Science ideas such as 'advocacy coalitions' and 'policy entrepreneurs') (Kern 2011), 'technology advocates' who do socio-political work to empower socio-technical niches, including by constructing actor-networks (Smith and Raven 2012) and 'cosmopolitan actors' who do socio-cognitive work to render technologies more widely applicable (Deuten 2003; Geels and Deuten 2006). But significant work needs to be done to develop these conceptualizations into a comprehensive theory that explicitly deals with the nature and role of such actors and, importantly, the political nature of their actions. Moreover, this needs to be situated within a systemic and active perspective of how transformations derive from such actions.

Implicit in all of the above are also critical questions of democracy and legitimacy. How do we ensure that the interests of poor people are represented in the governance of Socio-Technical Innovation System Building processes? What can we draw upon in order to inform such thinking? There are plenty of clues out there – for example, in the literature on reflexivity in the governance of technology and innovation and the now broad policy studies literatures on deliberative, discursive and participatory democracy. There is not space here to expand on these literatures, but the following references provide some good starting points (Dryzek 1990; Fischer and Forester 1993; Wynne 1993; Jasanoff 1999; Dryzek 2000; Fischer 2000; Wynne 2002; Fischer 2003; Stirling 2006; Smith and Stirling 2007; Ockwell 2008; Stirling 2008; Stirling 2009; Stirling 2011). In particular, recent thinking by our colleague Andy Stirling (Stirling 2015b) on the idea of 'innovation democracy' seems to offer a particularly fruitful avenue for approaching this. More explicit recent thinking on the politics of green transformations also offers several important insights in terms of the ways in which we might categorise and analyse such issues (Scoones et al. 2015a), as do various recent contributions on the politics of sustainable energy transitions (e.g. Baker et al. 2014). But plenty of work remains to be done on both the governance and politics of pro-poor green transformations.

Conclusion

Through detailed historical analysis of one of the most oft-cited examples of a transformation in sustainable energy access, this book has sought to move beyond the two-dimensional technology/engineering and finance/economics focus of the majority of the existing literature on sustainable energy access. We hope we have done enough to convince researchers and policy-makers of the urgent need to adopt a more systemic approach to understanding and acting on the problem of sustainable energy access, a perspective that is fully cognisant of how innovation and technological change have really happened in the past and of the socio-technical nature of successful examples of sustainable energy access. It is time to leave behind the current fixation on hardware financing and private sector entrepreneurship. Such interventions have their place, but they will never in and of themselves lead to the kinds of pro-poor green transformations that policy ambitions like SE4All and the SDGs imply. In the words of Ory Okolloh, a woman recognised as one of Kenya's leading technologists (and former policy lead for Google Africa), it is time to move beyond:

the fetishization around entrepreneurship in Africa. It's almost like, it's the next new liberal thing. Like, don't worry that there's no power, because, hey, you're going to do solar and innovate around that. Your schools suck, but, hey, there's this new model of schooling. Your roads are terrible, but, hey, Uber works in Nairobi and that's innovation.

During the Greek bail-out, no one was telling young Greek people to go and be entrepreneurs. Europe has been stuck at 2% or 1% growth. I don't see any entrepreneurship summit in Europe telling them, you know, go out there and be entrepreneurs. I feel that there's a sense that, oh, resilience and, you know, innovate around things – it's distracting us from dealing with fundamental problems that we cannot develop.[1]

The evidence presented in this book suggests that innovation does play an important role in transformative technological change and the development of markets for pro-poor green technologies, but it is not the kind of innovation that is addressed by venture capital and hardware financing. It is a systemic kind of innovation – one that involves the building of socio-technical innovation systems, changing the landscape within which sustainable technologies are adopted and built via the tireless and, doubtless, political efforts of key actors over time. The analysis in this book suggests that the success of the many laudable policy ambitions around sustainability, poverty alleviation and development more broadly rely fundamentally on a move towards this more systemic, socio-technical understanding of how pro-poor green transformations can be achieved. Clearly, there is still a long way to go. But we hope this book will make a small contribution to moving things in a pro-poor, green direction. The lives of many might be much better as a result.

Note

1 Quoted in an interview at the Quartz Africa Innovation Summit, 14 September 2015. Available at: http://qz.com/502149/video-ory-okolloh-explains-why-africa-cant-entrepreneur-itself-out-of-its-basic-problems/

REFERENCES

Abdel Latif, A. (2012) 'The UNEP-EPO-ICTSD project on patents and clean energy: A partnership to better understand the role of intellectual property rights in the transfer of climate friendly technologies', in Ockwell, D. G. and Mallett, A. (eds) *Low Carbon Technology Transfer: From Rhetoric to Reality*, London: Routledge.

Abdulla, A. (2008) Interview, Director of Telesales Solar, 14 July 2008, Nairobi.

Acker, R. H. and Kammen, D. M. (1996) 'The quiet (energy) revolution: Analysing the dissemination of photovoltaic power systems in Kenya', *Energy Policy*, 24(1), 81–111.

AEEP (n.d.) *The Energy Challenge: Access and Security for Africa and for Europe*, Eschborn: European Union Energy Initiative Partnership Dialogue Facility (EUEI PDF) on behalf of the African and European Implementing Teams of the Africa-EU Energy Partnership (AEEP).

AfDB (2013) *African Development Report 2012: Towards Green Growth in Africa*, Tunis: African Development Bank Group.

African Development Indicators (2009) *Africa Development Indicators (ADI), January 2009*, Manchester: ESDS International, (Mimas) University of Manchester.

African Union (2014) *On the Wings of Innovation: Science, Technology and Innovation for Africa 2024 STRATEGY (STISA-2024)*, Addis Ababa: African Union.

Ahlborg, H. and Sjöstedt, M. (2015) 'Small-scale hydropower in Africa: Socio-technical designs for renewable energy in Tanzanian villages', *Energy Research & Social Science*, 5, 20–33.

Akiyama, T. (1987) *Kenyan Coffee Sector Outlook: A Framework for Policy Analysis*, Commodity Studies and Projections Division, Washington, DC: The World Bank.

Anonymous (2013a) Interview, Government Regulator, 13 December 2013, Nairobi.

Anonymous (2013b) Interview, Multilateral Development Bank, 20 August 2013, Nairobi.

Arnold, E. and Bell, M. (2001) *Some New Ideas about Research for Development*, Copenhagen: Commission on Development-Related Research Funded by Danida.

Arungu-Olende, S. (2008) Interview, Former Coordinator for UN Conference on New and Renewable Sources of Energy, The African Academy of Sciences, 28 February 2008, Nairobi.

Avelino, F. and Rotmans, J. (2009) 'Power in transition: An interdisciplinary framework to study power in relation to structural change', *European Journal of Social Theory*, 12(4), 543–569.

Ayas, K. S. (1996) 'Design for learning for innovation', PhD thesis, Erasmus University.

Baker, L., Newell, P. and Phillips, J. (2014) 'The political economy of energy transitions: The case of South Africa', *New Political Economy*, 19(6), 791–818.

Barkan, J. (1994) 'Divergence and convergence in Kenya and Tanzania: Pressures for reform', in Barkan, J. (ed.) *Beyond Capitalism vs. Socialism in Kenya & Tanzania*, Nairobi and Boulder, CO: East African Educational Publishers and Lynne Rienner Publishers, pp. 1–45.

Barkan, J. D. and Holmquist, F. (1989) 'Peasant-state relations and the social base of self-help in Kenya', *World Politics*, 41(3), 359–380.

Barnett, A. (1990) 'The diffusion of energy technology in the rural areas of developing countries: A synthesis of recent experience', *World Development*, 18(4), 539–553.

Bazilian, M. and Pielke, R. J. R. (2013) 'Making energy access meaningful', *Issues in Science and Technology*, 29(4), 74–78.

BCEOM, EAA and FONDEM (2001) *Study on Solar Photovoltaics (PV) Quality and Service Specification and Market Penetration*, Final Report, Nairobi: Ministry of Energy.

Bell, M. (1990) *Continuing Industrialisation, Climate Change and International Technology Transfer*, Oslo and Brighton: Resource Policy Group and Science Policy Research Unit.

Bell, M. (1997) 'Technology transfer to transition countries: Are there lessons from the experience of the post-war industrializing countries?', in Dyker, D. (ed.) *The Technology of Transition: Science and Technology Policies for Transition Countries*, Budapest: Central European University Press, pp. 63–94.

Bell, M. (2009) 'Innovation Capabilities and Directions of Development', *STEPS Working Paper*, 33, Brighton: STEPS Centre.

Bell, M. (2012) 'International technology transfer, innovation capabilities and sustainable directions of development', in Ockwell, D. G. and Mallett, A. (eds) *Low-Carbon Technology Transfer: From Rhetoric to Reality*, London: Routledge, pp. 20–47.

Bell, M. and Pavitt, K. (1993) 'Technological accumulation and industrial growth: Contrasts between developed and developing countries', *Industrial and Corporate Change*, 2(2), 157–210.

Berkhout, F. (2006) 'Normative expectations in systems innovation', *Technology Analysis & Strategic Management*, 18(3–4), 299–311.

Berkhout, F., Smith, A. and Stirling, A. (2004) 'Socio-technological regimes and transition contexts', in Elzen, B., Geels, F. and Green, K. (eds) *System Innovation and the Transitions to Sustainability: Theory, Evidence and Policy*, Cheltenham: Edward Elgar, pp. 48–75.

Berkhout, F., Verbong, G., Wieczorek, A. J., Raven, R., Lebel, L. and Bai, X. (2010) 'Sustainability experiments in Asia: Innovations shaping alternative development pathways?', *Environmental Science and Policy*, 13(4), 261–271.

Bess, M. (2002) 'Sustainable energy in Kenya: An overview – the Good, the Bad & the…', *Policy Dialogue and Sustainable Energy in Kenya*, Nairobi, 21 May.

Bevan, D., Collier, P. and Gunning, J. (1987) 'Consequences of a commodity boom in a controlled economy: Accumulation and redistribution in Kenya, 1975–83', *The World Bank Economic Review*, 1(3), 489–513.

Bevan, D., Collier, P. and Gunning, J. (1990) 'Fiscal response to a temporary shock: The aftermath of the Kenyan coffee boom', *The World Bank Economic Review*, 3(3), 359–378.

Blyth, L. (2008) Interview, Engineer/Entrepreneur, Sunpak, 18 July 2008, Nairobi.

Bresson, R. (2001) 'PVMTI, come rain or shine we're carrying on regardless', *SolarNet Magazine*, 3(2), 5–6.

Brewer, T. (2008) 'Climate change technology transfer: A new paradigm and policy agenda', *Climate Policy*, 8, 516–526.

Breznitz, D. and Murphree, M. (2011) *Run of the Red Queen: Government, Innovation, Globalization, and Economic Growth in China*, New Haven, CT: Yale University Press.

Byrne, R. (1999) 'Between the spear and the book: Solar PV in Maasailand, Tanzania', *SolarNet Magazine*, 1(2), 13–14.

Byrne, R. (2011) 'Learning drivers: rural electrification regime building in Kenya and Tanzania', DPhil, SPRU, University of Sussex.

Byrne, R. (2013a) *Climate Technology and Development Case Study: Compact Fluorescent Lamps (CFLs)*, Cambridge: Climate Strategies.

Byrne, R. (2013b) 'Low carbon development in Tanzania: Lessons from its solar home system market', in Urban, F. and Nordensvärd, J. (eds) *Low Carbon Development: Key Issues*, London: Earthscan, pp. 240–255.

Byrne, R., de Coninck, H. and Sagar, A. (2014a) *Low Carbon Innovation for Industrial Sectors in Developing Countries*, Cambridge: Climate Strategies.

Byrne, R., Ockwell, D., Urama, K. et al. (2014b) 'Sustainable energy for whom? Governing pro-poor, low carbon pathways to development: Lessons from solar PV in Kenya', *STEPS Working Paper*, 61, Brighton: STEPS Centre.

Byrne, R., Schoots, K., Watson, J., Ockwell, D., Gallagher, K. and Sagar, A. (2012a) *Innovation Systems in Developing Countries*, Cambridge: Climate Strategies.

Byrne, R., Smith, A., Watson, J. and Ockwell, D. (2012b) 'Energy pathways in low carbon development: The need to go beyond technology transfer', in Ockwell, D. and Mallett, A. (eds) *Low Carbon Technology Transfer: From Rhetoric to Reality*, London: Routledge, pp. 123–142.

Caniëls, M. C. J. and Romijn, H. A. (2008) 'Strategic niche management: Towards a policy tool for sustainable development', *Technology Analysis & Strategic Management*, 20(2), 245–266.

Chaminade, C., Lundvall, B.-Å., Vang, J. and Joseph, K. (2009) 'Designing innovation policies for development: towards a systemic experimentation-based approach', in Lundvall, B.-Å., et al. (eds) *Handbook of Innovation Systems and Developing Countries: Building Domestic Capabilities in a Global Setting*, Cheltenham: Edward Elgar, pp. 360–379.

Chang, H.-J. (2002) *Kicking Away the Ladder: Development Strategy in Historical Perspective*, London: Anthem Press.

Chattopadhyay, D., Bazilian, M. and Lilienthal, P. (2015) 'More power, less cost: Transitioning up the solar energy ladder from home systems to mini-grids', *The Electricity Journal*, 28(3), 41–50.

Cimoli, M., Dosi, G. and Stiglitz, J. (eds) (2009) *Industrial Policy and Development: The Political Economy of Capabilities Accumulation*, New York: Oxford University Press.

Cohen, W. M. and Levinthal, D. A. (1990) 'Absorptive capacity: A new perspective on learning and innovation', *Administrative Science Quarterly*, 35(1), 128–152.

Conway, S. and Steward, F. (2009) *Managing and Shaping Innovation*, New York: Oxford University Press.

Cooke, P., Gomez Uranga, M. and Etxebarria, G. (1997) 'Regional innovation systems: Institutional and organisational dimensions', *Research Policy*, 26(4–5), 475–491.

Cornell University, INSEAD and WIPO (2015) *The Global Innovation Index 2015: Effective Innovation Policies for Development*, Geneva: World Intellectual Property Organization.

CTCN (2014) *CTCN Operating Manual for National Designated Entities (NDEs)*, Copenhagen: Climate Technology Centre and Network.

D'Costa, A. P. (1998) 'Coping with technology divergence policies and strategies for India's industrial development', *Technological Forecasting and Social Change*, 58(3), 271–283.

de Bakker, P. (2001) 'PVMTI: Is the old "green carrot" becoming a pie in the sky?', *SolarNet Magazine*, 3(2), 4–5.

de Coninck, H. and Puig, D. (2015) 'Assessing climate change mitigation technology interventions by international institutions', *Climatic Change*, 131(3), 417–433.

Delucchi, M. A. and Jacobson, M. Z. (2011) 'Providing all global energy with wind, water, and solar power, Part II: Reliability, system and transmission costs, and policies', *Energy Policy*, 39(3), 1170–1190.

Delucchi, M. A. and Jacobson, M. Z. (2012) 'Response to "A critique of Jacobson and Delucchi's proposals for a world renewable energy supply" by Ted Trainer', *Energy Policy*, 44, 482–484.

Deuten, J. J. (2003) 'Cosmopolitanising technologies: A study of four emerging technological regimes', PhD thesis, University of Twente.

Deuten, J., Rip, A. and Jelsma, J. (1997) 'Societal embedding and product creation management', *Technology Analysis & Strategic Management*, 9(2), 131–148.

Development Marketplace (2007) *Innovations in Off-Grid Lighting Products and Services for Africa* Lighting Africa Development Marketplace Grant Competition Guidelines, Washington, DC: Lighting Africa Development Marketplace.

Disenyana, T. (2009) *China in the African Solar Energy Sector: Kenya Case Study*, Johannesburg: South African Institute of International Affairs.

Doranova, A. (2010) *Technology Transfer and Learning Under the Kyoto Regime: Exploring the Technological Impact of CDM Projects in Developing Countries*, Maastricht: UNU-MERIT, University of Maastricht.

Dosi, G. (1982) 'Technological paradigms and technological trajectories: a suggested interpretation of the determinants and directions of technical change', *Research Policy*, 11, 147–162.

Dosi, G. (1988) 'The nature of the innovative process', in Dosi, G., et al. (eds) *Technical Change and Economic Theory*, London: Pinter.

Dosi, G. and Nelson, R. (1993) 'Evolutionary Theories in Economics: Assessment and Prospects', *Working Paper*, WP-93-064, Vienna: International Institute for Applied Systems Analysis.

Douthwaite, B. and Ashby, J. (2005) *Innovation Histories: A Method for Learning from Experience*, Fiumicino: Institutional Learning and Change (ILAC) Initiative.

Dryzek, J. S. (1990) *Discursive Democracy: Politics, Policy and Political Science*, New York: Cambridge University Press.

Dryzek, J. S. (2000) *Deliberative Democracy and Beyond: Liberals, Critics, Contestations*, Oxford: Oxford University Press.

Duffy, J., Campbell, H., Sajo, A. and Sanz, E. (1988) *Photovoltaic Systems for Export Application*, Cambridge, MA: Massachusetts Institute of Technology Energy Laboratory.

Duke, R., Graham, S., Hankins, M., et al. (2000) *Field Performance Evaluation of Amorphous Silicon (a-Si) Photovoltaic Systems in Kenya: Methods and Measurements in Support of a Sustainable Commercial Solar Energy Industry*, Washington, DC: UNDP/World Bank.

Duke, R. D., Jacobson, A. and Kammen, D. M. (2002) 'Photovoltaic module quality in the Kenyan solar home systems market', *Energy Policy*, 30(6), 477–499.

EAA (1998) *The Kenyan Solar Photovoltaic Industry: Industry Status and Key Players*, Nairobi: PVMTI.

EAA (2001) *The Kenya Portable Battery Pack Experience: Test Marketing an Alternative for Low-income Rural Household Electrification*, Washington, DC: UNDP/World Bank ESMAP.

Eames, M., McDowall, W., Hodson, M. and Marvin, S. (2006) 'Negotiating contested visions and place-specific expectations of the hydrogen economy', *Technology Analysis & Strategic Management*, 18(3/4), 361–374.

ERC (2012) *The Energy (Solar Photovoltaic Systems) Regulations 2012*, Nairobi: Energy Regulatory Commission, Republic of Kenya.

ERC (2014a) *Register of Licensed Solar Photovoltaic Contractors as at January 2014, ERC/SPV/ 3.01*, Nairobi: Energy Regulatory Commission.

ERC (2014b) *Updated Register of Licensed Solar Photovoltaic Technicians as at January 2014, ERC/SPV/3.01*, Nairobi: Energy Regulatory Commission.

ERC (2015a) *Solar Photovoltaic Contractors Register as at 04/11/2015, ERC/SPV/3.01*, Nairobi: Energy Regulatory Commission.

ERC (2015b) *Solar Photovoltaic Technicians Register as at 04 November 2015, ERC/SPV/3.01*, Nairobi: Energy Regulatory Commission.

Escobar, A. (2012) *Encountering Development: The Making and Unmaking of the Third World*, Paperback reissue edn, Princeton, NJ: Princeton University Press.

ESD (2003) *Study on PV Market Chains in East Africa (Draft Final Copy), Report for the World Bank*, Nairobi: Energy for Sustainable Development.

Fagerberg, J. (2005) 'Innovation: A guide to the literature', in Fagerberg, J., Mowery, D. and Nelson, R. (eds) *The Oxford Handbook of Innovation*, New York: Oxford University Press, pp. 1–26.

Fagerberg, J. and Godinho, M. (2005) 'Innovation and catching-up', in Fagerberg, J., Mowery, D. and Nelson, R. (eds) *The Oxford Handbook of Innovation*, New York: Oxford University Press, pp. 514–542.

Fairhead, J. and Leach, M. (1996) *Misreading the African Landscape*, Cambridge: Cambridge University Press.

Fischer, F. (2000) *Citizens, Experts and the Environment*, Durham, NC: Duke University Press.

Fischer, F. (2003) *Reframing Public Policy. Discursive Politics and Deliberative Practices*, Oxford: Oxford University Press.

Fischer, F. and Forester, J. (1993) *The Argumentative Turn in Policy Analysis and Planning*, Durham, NC: Duke University Press.

Foley, G. (1995) *Photovoltaic Applications in Rural Areas of the Developing World*, Washington, DC: Energy Sector Management Assistance Programme.

Forsyth, T. (2005) 'Enhancing climate technology transfer through greater public-private cooperation: lessons from Thailand and the Philippines', *Natural Resource Forum*, 29 (2), 165–176.

Forsyth, T. (2007) 'Promoting the "development dividend" of climate technology transfer: can cross-sector partnerships help?', *World Development*, 35(10), 1684–1698.

Forsyth, T. (2008) 'Political ecology and the epistemology of social justice', *Geoforum*, 39 (2), 756–764.

Foster, C. and Heeks, R. (2013) 'Conceptualising inclusive innovation: Modifying systems of innovation frameworks to understand diffusion of new technology to low-income consumers', *The European Journal of Development Research*, 25(3), 333–355.

Freeman, C. (1987) *Technology and Economic Performance: Lessons from Japan*, London: Pinter.

Freeman, C. (1992) *The Economics of Hope: Essays on Technical Change, Economic Growth and the Environment*, London: Pinter.

Freeman, C. (1997) 'The national system of innovation in historical perspective', *Cambridge Journal of Economics*, 19, 5–24.

Freeman, C. (2002) 'Continental, national and sub-national innovation systems – complementarity and economic growth', *Research Policy*, 31(2), 191–211.

Fressoli, M., Arond, E., Abrol, D., et al. (2014) 'When grassroots innovation movements encounter mainstream institutions: Implications for models of inclusive innovation', *Innovation and Development*, 4(2), 277–292.

Furlong, K. (2014) 'STS beyond the "modern infrastructure ideal": extending theory by engaging with infrastructure challenges in the South', *Technology in Society*, 38, 139–147.

Gallagher, K. S. (2006) 'Limits to leapfrogging in energy technologies? Evidence from the Chinese automobile industry', *Energy Policy*, 34(4), 383–394.

Geels, F. (2002) 'Technological transitions as evolutionary reconfiguration processes: A multi-level perspective and a case-study', *Research Policy*, 31, 1257–1274.

Geels, F. (2004) 'From sectoral systems of innovation to socio-technical systems: Insights about dynamics and change from sociology and institutional theory', *Research Policy*, 33, 897–920.

Geels, F. (2014) 'Regime resistance against low-carbon transitions: Introducing politics and power into the multi-level perspective', *Theory, Culture & Society*, 31(5), 21–40.

Geels, F. and Deuten, J. J. (2006) 'Local and global dynamics in technological development: A socio-cognitive perspective on knowledge flows and lessons from reinforced concrete', *Science and Public Policy*, 33(4), 265–275.

Geels, F. and Raven, R. (2006) 'Non-linearity and expectations in niche-development trajectories: Ups and downs in Dutch biogas development (1973–2003)', *Technology Analysis & Strategic Management*, 18(3–4), 375–392.

Geels, F. W. and Schot, J. (2007) 'Typology of sociotechnical transition pathways', *Research Policy*, 36(3), 399–417.

Gisore, G. (2002) 'Codes and standards: codes of practice and installation standards', in EAA and ERG (eds) *Solar Market Development and Capacity Building in Kenya: The Role of Technicians and Upcountry Vendors in the Kenyan Solar Industry*, Nairobi: Energy Alternatives Africa (Nairobi) and Energy and Resources Group (University of California, Berkeley), Methodist Guest House, pp. 43–53.

Giuliani, E. and Bell, M. (2005) 'The micro-determinants of meso-level learning and innovation: Evidence from a Chilean wine cluster', *Research Policy*, 34(1), 47–68.

GOGLA and Lighting Global (2015) 'Conference Report: 4th International Off-Grid Lighting Conference and Exhibition', October 26–29, Dubai, United Arab Emirates, Global Off-Grid Lighting Association and Lighting Global.

Gollwitzer, L., Ockwell, D. G. and Ely, A. (2015) 'Institutional innovation in the management of pro-poor energy access in East Africa', *SPRU Working Paper Series*, SWPS 2015–29, Brighton: SPRU.

Goodman, G. (1984) 'Foreword', in O'Keefe, P., Raskin, P. and Bernow, S. (eds) *Energy and Development in Kenya: Opportunities and Constraints*, Stockholm and Uppsala: Beijer Institute and the Scandinavian Institute of African Studies.

Grin, J. (2010) 'Understanding transitions from a governance perspective', in Grin, J. et al. (eds) *Transitions to Sustainable Development: New Directions in the Study of Long Term Transformative Change*, London: Routledge, pp. 221–319.

Gunning, R. (2003) 'The Photovoltaic Market Transformation Initiative', in IEA (ed.) *16 Case Studies on the Deployment of Photovoltaics Technologies in Developing Countries, IEA-PVPS T9–07:2003*, Paris: International Energy Agency, pp. 81–89.

Hankins, M. (1987) *Renewable Energy in Kenya*, Nairobi: Motif Creative Arts.

Hankins, M. (1990) *Optimising performance of small solar electric systems in rural Kenya: Technical and social approaches*, MSc dissertation, University of Reading.

Hankins, M. (1993) *Solar Rural Electrification in the Developing World. Four Country Case Studies: Dominican Republic, Kenya, Sri Lanka and Zimbabwe*, Washington, DC: Solar Electric Light Fund.

Hankins, M. (1995) *Solar Electric Systems for Africa: A Guide for Planning and Installing Solar Electric Systems in Rural Africa*, Revised ed, Harare: AGROTEC and Commonwealth Science Council.

Hankins, M. (1996) *Lighting Services for the Rural Poor: Test Marketing and Evaluation of 7 Solar Lantern Units in Rural Kenya*, Washington, DC: World Bank.

Hankins, M. (1999) 'Ten years of solar PV training in East Africa', *SolarNet Magazine*, 1(2), 6–8.

Hankins, M. (2003) 'The Kenya PV experience', draft background paper, Workshop on Financing Mechanisms and Business Models for PV Systems in Africa, Pretoria, South Africa, 27–29 May.

Hankins, M. (2007) Interview, Former Managing Director, EAA, 16 November 2007, Nairobi.

Hankins, M. and Bess, M. (1994) *Photovoltaic Power to the People: The Kenya Case*, Washington, DC: UNDP-World Bank Energy Sector Management Assistance Programme.

Hankins, M., Ochieng, F. and Scherpenzeel, J. (1997) *PV Electrification in Rural Kenya: A Survey of 410 Solar Home Systems in 12 Districts, Final Report*, Washington, DC: UNDP-World Bank Energy Sector Management Assistance Programme.

Hankins, M., Saini, A. and Kirai, P. (2009) *Uganda's Solar Energy Market: Target Market Analysis*, Berlin: GTZ.

Hankins, M. and van der Plas, R. (2000) *Implementation Manual: Financing Mechanisms for Solar Electric Equipment, Report ESM231/00*, Washington, DC: UNDP-World Bank Energy Sector Management Assistance Programme.

Hansen, U. E. and Nygaard, I. (2014) 'Sustainable energy transitions in emerging economies: The formation of a palm oil biomass waste-to-energy niche in Malaysia 1990–2011', *Energy Policy*, 66, 666–676.

Hansen, U. E. and Ockwell, D. (2014) 'Learning and technological capability building in emerging economies: The case of the biomass power equipment industry in Malaysia', *Technovation*, 34(10), 617–630.

Hansen, U. E., Pedersen, M. B. and Nygaard, I. (2015) 'Review of solar PV policies, interventions and diffusion in East Africa', *Renewable and Sustainable Energy Reviews*, 46, 236–248.

Haum, R. (2012) 'Project based market transformation in developing countries and international technology transfer: The case of the Global Environment Facility and Solar PV', in Ockwell, D. and Mallett, A. (eds) *Low Carbon Technology Transfer: From Rhetoric to Reality*, London: Routledge, pp. 185–208.

Hekkert, M., Suurs, R., Negro, S., Kuhlmann, S. and Smits, R. (2007) 'Functions of innovation systems: A new approach for analysing technological change', *Technological Forecasting and Social Change*, 74, 413–432.

Henderson, R. (1989) 'World Health Organization Expanded Programme on Immunization: Progress and Evaluation Report', *Annals of the New York Academy of Sciences*, 569, 45–68.

Hobday, M. (1995a) 'East Asian latecomer firms: Learning the technology of electronics', *World Development*, 23(7), 1171–1193.

Hobday, M. (1995b) *Innovation in East Asia: The Challenge to Japan*, Aldershot: Edward Elgar.

Hoogma, R. (2000) 'Exploiting technological niches', PhD thesis, University of Twente.

Hoogma, R., Kemp, R., Schot, J. and Truffer, B. (2002) *Experimenting for Sustainable Transport: The Approach of Strategic Niche Management*, London: Spon Press.

Howells, J. (2006) 'Intermediation and the role of intermediaries in innovation', *Research Policy*, 35(5), 715–728.

Hulme, M. (2008) 'Governing and adapting to climate. A response to Ian Bailey's Commentary on "Geographical work at the boundaries of climate change"', *Transactions of the Institute of British Geographers*, 33(3), 424–427.

Hulme, M. (2009) *Why We Disagree About Climate Change*, Cambridge: Cambridge University Press.

ICCEPT and E4tech (2003) *The UK Innovation Systems for New and Renewable Energy Technologies: Final Report*, London: DTI Renewable Energy Development and Deployment Team.

IDRC (2011) *Innovation for Inclusive Development: Program Prospectus for 2011–2016*, Ottawa: IDRC.

IEA (2014) *Africa Energy Outlook*, Paris: International Energy Agency.

IFC (1998) *India, Kenya, and Morocco: Photovoltaic Market Transformation Initiative (PVMTI), Project Document*, Washington, DC: International Finance Corporation.

IFC (2007a) *Lighting the Bottom of the Pyramid, GEF Project Appraisal Document*, Washington, DC: International Finance Corporation.

IFC (2007b) *Selling Solar: Lessons from More Than a Decade of the IFC's Experience*, Washington, DC: International Finance Corporation.

infoDev (2015) *The Climate Technology Programme: Accelerating Climate Innovation in Developing Countries*, Washington, DC: World Bank and International Finance Corporation.

IRENA (2015) *Renewable Power Generation Costs in 2014*, International Renewable Energy Agency.

Jacobson, A. (2002a) 'Solar technicians & capacity building in Kenya: Results from the Solar Technician Evaluation Project', Workshop on Solar Market Development and Capacity Building in Kenya: The Role of Technicians and Upcountry Vendors in the Kenyan Solar Industry, 21 August, Nairobi: Methodist Guest House, pp. 6–19.

Jacobson, A. (2002b) 'Solar technicians and upcountry vendors: Key trends in the Kenyan solar supply chain', Workshop on Solar Market Development and Capacity Building in Kenya: The Role of Technicians and Upcountry Vendors in the Kenyan Solar Industry, 21 August, Nairobi: Methodist Guest House, pp. 29–40.

Jacobson, A. (2004) 'Connective power: Solar electrification and social change in Kenya', PhD degree, University of California.

Jacobson, A. (2007) 'Connective power: Solar electrification and social change in Kenya', *World Development*, 35(1), 144–162.

Jacobson, A. and Kammen, D. (2005) *The Value of Vigilance: Evaluating Product Quality in the Kenyan Photovoltaics Industry*, Berkeley, CA: Humboldt State University and University of California Berkeley,

Jacobson, M. Z. and Delucchi, M. A. (2011) 'Providing all global energy with wind, water, and solar power, Part I: Technologies, energy resources, quantities and areas of infrastructure, and materials', *Energy Policy*, 39(3), 1154–1169.

Jacobsson, S., Sandén, B. and Bångens, L. (2004) 'Transforming the energy system: The evolution of the German technological system for solar cells', *Technology Analysis & Strategic Management*, 16(1), 3–30.

Jasanoff, S. (1999) 'STS and public policy: Getting beyond deconstruction', *Science Technology and Society*, 4(1), 59–72.

Johnson, F. (1939) *A Standard Swahili-English Dictionary, Founded on Madan's Swahili-English Dictionary by the Inter-Territorial Language Committee for the East African Dependencies under the Direction of Frederick Johnson*, Nairobi: Oxford University Press.

Jones, M. (1986) *Energy Conservation in Kenya: Experiences of the Kenya Renewable Energy Development Project*, Washington, DC and Nairobi: Energy/Development International for the Ministry of Energy and Regional Development, Republic of Kenya.

Karekezi, S. (1994) 'Disseminating renewable energy technologies in sub-Saharan Africa', *Annual Review of Energy and the Environment*, 19(1), 387–421.

Katz, J. (1987) *Technology Generation in Latin American Manufacturing Industries*, London: Macmillan.

Keane, J. (2005) 'Locally-made solar panels for small appliances', *Boiling Point*, 51, 7.

KEBS (2014) *Approved List of Standards by 103rd Standards Approval Committee Meeting on 19th June 2014*, Nairobi: Kenya Bureau of Standards.

Keeley, J. and Scoones, I. (2003) *Understanding Environmental Policy Processes. Cases from Africa*, London: Earthscan.

Kemp, R., Schot, J. and Hoogma, R. (1998) 'Regime shifts to sustainability through processes of niche formation: The approach of strategic niche management', *Technology Analysis & Strategic Management*, 10(2), 175–195.

KEREA (2005) *Code of Conduct*, Nairobi: Kenya Renewable Energy Association.

KEREA (2012) 'Solar PV Curriculum Development', workshop report, Kenya Renewable Energy Association and UNDP, Elsmere Conservation Center, Naivasha, 9–13 July 2012.

Kern, F. (2011) 'Ideas, institutions, and interests: Explaining policy divergence in fostering "system innovations" towards sustainability', *Environment and Planning C: Government and Policy*, 29(6), 1116–1134.

Khan, M. and Blankenburg, S. (2009) 'The political economy of industrial policy in Asia and Latin America', in Cimoli, M., Dosi, G. and Stiglitz, J. (eds) *Industrial Policy and Development: The Political Economy of Capabilities Accumulation*, New York: Oxford University Press, pp. 336–377.

Kilonzo, A. (2013) Interview, Former Coordinator, SolarNet, 29 July 2013, Nairobi.

Kim, Y., Kim, L. and Lee, J. (1989) 'Innovation strategy of local pharmaceutical firms in Korea: A multivariate analysis', *Technology Analysis & Strategic Management*, 1(1), 29–44.

Kimani, M. (ed.) (1992) *Regional Solar Electric Training and Awareness Workshop, Proceedings of a Workshop held in Nairobi and Meru, 15–27 March 1992*, Washington, DC and Nairobi: African Development Foundation and M. Kimani.

Kimani, M. and Hankins, M. (1993) 'Rural PV lighting systems: A case study of indigenous demand-led technology uptake', in Walubengo, D. and Kimani, M. (eds) *Whose Technologies? The Development and Dissemination of Renewable Energy Technologies (RETs) in Sub-Saharan Africa*, Nairobi: KENGO Regional Wood Energy Programme for Africa (RWEPA).

Kimuya, J. (2013) Interview, Sales and Marketing Executive, Ubbink EA Ltd., 21 August 2013, Naivasha.

Kithokoi, D. (2008) Interview, Formerly of Solar Shamba, DAMUKI Enterprises Ltd., 11 July 2008, Nairobi.

Kivimaa, P. (2014) 'Government-affiliated intermediary organisations as actors in system-level transitions', *Research Policy*, 43(8), 1370–1380.

Kline, S. and Rosenberg, N. (1986) 'An overview of innovation', in Landau, R. and Rosenberg, N. (eds) *The Positive Sum Strategy: Harnessing Technology for Economic Growth*, Washington, DC: National Academy Press, pp. 275–305.

Konrad, K. (2006) 'The social dynamics of expectations: The interaction of collective and actor-specific expectations on electronic commerce and interactive television', *Technology Analysis & Strategic Management*, 18(3/4), 429–444.

KPMG (2011) *Budget Brief Kenya 2011*, Nairobi: KPMG Kenya.

KSTF (2009) 'KARADEA Solar Training Facility pages of the KARADEA website' [online]. Available at: www.karadea.8k.com/projects.htm [accessed 24 February 2014].

Kuo, L. (2015) 'Video: Ory Okolloh explains why Africa can't entrepreneur itself out of its basic problems', Quartz Africa Innovation Summit, 14 September 2015. Available at: http://qz.com/502149/video-ory-okolloh-explains-why-africa-cant-entrepreneur-itself-out-of-its-basic-problems/ [accessed 8 July 2016].

Lawhon, M. and Murphy, J. T. (2012) 'Socio-technical regimes and sustainability transitions: Insights from political ecology', *Progress in Human Geography*, 36(3), 354–378.

Leach, M. and Mearns, R. (1996) *The Lie of the Land. Challenging Received Wisdom on the African Environment*, Oxford: James Curry.

Leach, M., Rockström, J., Raskin, P., et al. (2012) 'Transforming innovation for sustainability', *Ecology and Society*, 17(2), 11.

Leach, M., Scoones, I. and Stirling, A. (2010a) *Dynamic Sustainabilities: Technology, Environment, Social Justice*, London: Routledge.

Leach, M., Scoones, I. and Stirling, A. (2010b) 'Governing epidemics in an age of complexity: Narratives, politics and pathways to sustainability', *Global Environmental Change*, 20(3), 369–377.

Lecocq, F. and Ambrosi, P. (2007) 'The Clean Development Mechanism: History, status, and prospects', *Review of Environmental Economics and Policy*, 1(1), 134–151.

Lehtonen, M. (2011) 'Social sustainability of the Brazilian bioethanol: Power relations in a centre-periphery perspective', *Biomass and Bioenergy*, 35(6), 2425–2434.

Lema, A. and Lema, R. (2013) 'Technology transfer in the Clean Development Mechanism: Insights from wind power', *Global Environmental Change: Human and Policy Dimensions*, 23(1), 301–313.

Lighting Africa (2008a) *Kenya: Qualitative Off-Grid Lighting Market Assessment*, Washington, DC: International Finance Corporation.

Lighting Africa (2008b) *Lighting Africa Market Assessment Results, Quantitative Assessment: Kenya*, Washington, DC: International Finance Corporation.

Lighting Africa (2008c) *Lighting Africa Year 1: Progress and Plans, Annual Report*, Washington, DC: International Finance Corporation.

Lighting Africa (2009) *Lighting Africa Year 2: Progress Update*, Washington, DC: International Finance Corporation.

Lighting Africa (2011) *Second International Business Conference and Trade Fair, Nairobi 18–20 May 2010, Conference Report*, Washington, DC: World Bank and International Finance Corporation.

Lighting Africa (2013a) '3rd International Off-Grid Lighting Conference and Trade Fair', November 13–15, 2012, Conference Proceedings, King Fahd Hotel Dakar, Senegal: Lighting Africa.

Lighting Africa (2013b) 'Consumer education campaign scoops Kenya marketing award', *Lighting Africa Newsletter*, p. 1.

Lighting Africa (2014) 'Program Results as of December 2014'. Available at: www.lighting africa.org/wp-content/uploads/2014/07/Lighting-Africa-Results_Dec2014_Final.pdf [accessed 15 December 2015].

Loh, V. (2007) Interview, Chairman, KEREA – Kenya Renewable Energy Association, 15 November 2007, Nairobi.

Lundvall, B.-Å. (1988) 'Innovation as an interactive process: From user-producer interaction to the national system of innovation', in Dosi, G., *et al.* (eds) *Technical Change and Economic Theory*, London: Pinter, pp. 349–369.

Lundvall, B.-Å. (1992) *National Systems of Innovation: Towards a Theory of Innovation and Interactive Learning*, London: Pinter.

Mabonga, P. S. (2013) Interview, Sales and Service Engineer (Chloride Exide), Technical Director (Altech Engineering), 19 August and 25 November 2013, Nairobi.

Magambo, K. (2006) *Curriculum for the Training and Certification of Photovoltaic Practitioners. Basic PV Training Courses: Syllabi and Regulations*, Nairobi: Kenya Renewable Energy Association.

Malerba, F. and Mani, S. (eds) (2009) *Sectoral Systems of Innovation and Production in Developing Countries: Actors, Structure and Evolution*, Cheltenham: Edward Elgar.

Markard, J., Raven, R. and Truffer, B. (2012) 'Sustainability transitions: An emerging field of research and its prospects', *Research Policy*, 41(6), 955–967.

Marshall, M., Ockwell, D. and Byrne, R. (in press) 'Sustainable Energy for All, or Sustainable Energy for Men? Gender and the construction of identity within climate technology entrepreneurship in Kenya', *Progress in Development Studies*.

Marvin, S., Guy, S., Medd, W. and Moss, T. (2011) 'Conclusions: The transformative power of intermediaries', in Guy, S., et al. (eds) *Shaping Urban Infrastructures: Intermediaries and the Governance of Socio-Technical Networks*, London: Earthscan, pp. 209–217.

Masakhwe, J. (1993) 'The private sector in solar industry: The case of Kenya', in Kimani, M. and Naumann, E. (eds) *Seminar on Recent Experiences in Research, Development, and Dissemination of Renewable Energy Technologies in Sub-Saharan Africa*, 29 March–2 April, Nairobi: KENGO International Outreach Department, pp. 65–69.

Matsuo, N. (2003) 'CDM in the Kyoto negotiations: How CDM has worked as a bridge between developed and developing worlds', *Mitigation and Adaptation Strategies for Global Change*, 8, 191–200.

Mazzucato, M. (2013) *The Entrepreneurial State: Debunking Public vs. Private Sector Myths*, London: Anthem Press.

Mbithi, P. (2007) Case Study: Photovoltaics for Schools Project, Presentation given at ESDA Nairobi by Paul Mbithi of the Ministry of Energy.

Mboa, A. (2013) Interview, Standards Officer, KEBS, 20 August 2013, Nairobi.

McNelis, B., Derrick, A. and Starr, M. (1988) *Solar-Powered Electricity: A Survey of Photovoltaic Power in Developing Countries*, London: Intermediate Technology Publications in association with UNESCO.

Meadowcroft, J. (2011) 'Engaging with the politics of sustainability transitions', *Environmental Innovation and Societal Transitions*, 1(1), 70–75.

Meza, E. (2013) 'Special report Africa: Tanzania, Mozambique', *PV-magazine*. Available at: m.pv-magazine.com/news/details/beitrag/special-report-africa [accessed 8 July 2016].

Michael, M. (2000) 'Futures of the present: From performativity to prehension', in Brown, N., Rappert, B. and Webster, A. (eds) *Contested Futures: A Sociology of Prospective Techno-Science*, Aldershot: Ashgate Publishing, pp. 21–39.

Migai, F. (2008) Interview, Micro-solar entrepreneur, 7 February 2008, Nairobi.

Modi, V., McDade, S., Lallement, D. and Saghir, J. (2005) *Energy Services for the Millennium Development Goals*, Washington, DC: Energy Sector Management Assistance Programme, United Nations Development Programme, UN Millennium Project, and World Bank.

MOE (2012) *Feed-in-Tariffs Policy for Wind, Biomass, Small Hydros, Geothermal, Biogas and Solar*, second revision, Nairobi: Ministry of Energy, Republic of Kenya.

MOE (2013) *National Energy Policy: November 2013 Draft*, Nairobi: Ministry of Energy and Petroleum, Republic of Kenya.

Moussa, S. Z. (2002) 'Technology Transfer for Agricultural Growth in Africa', *Economic Research Paper* no. 72, Abidjan, Côte d'Ivoire: African Development Bank.

Muchiri, D. (2001) 'PVMTI and the Kenyan PV market', *SolarNet Magazine*, 3(2), 4.

Muchiri, D. (2008) Interview, Projects Manager, Solar World (EA) Ltd., 7 July 2008, Nairobi.

Mugalo, B. (1981) *Prospects for New and Renewable Sources of Energy in Kenya: Extracts from the Kenya National Paper Prepared for the United Nations Conference on New and Renewable Sources of Energy*, Nairobi: Ministry of Energy.

Mulugetta, Y. and Urban, F. (2010) 'Deliberating on low carbon development', *Energy Policy*, 38(12), 7546–7549.

Musinga, M., Hankins, M., Hirsch, D. and de Schutter, J. (1997) *Kenya Photo-Voltaic Rural Energy Project (KENPREP): Results of the 1997 Market Survey*, Nairobi: DGIS, Ecotec Ltd.

Mutimba, S. (2002a) *Fourth Policy Dialogue Meeting on Sustainable Energy in Kenya*, Nairobi: ESD.

Mutimba, S. (2002b) *Sixth Policy Dialogue Meeting on Sustainable Energy in Kenya Focusing on International Financing Mechanisms*, Nairobi: ESD.

Mutimba, S. (2002c) *Third Policy Dialogue Meeting on Sustainable Energy in Kenya*, Nairobi: ESD.

Mutimba, S. (2007) Interview, Managing Director, ESDA, 29 November 2007, Nairobi.

Naess, L. O., Newell, P., Newsham, A., Phillips, J., Quan, J. and Tanner, T. (2015) 'Climate policy meets national development contexts: Insights from Kenya and Mozambique', *Global Environmental Change*, 35, pp. 534–544.

Nelson, R. and Winter, S. (1982) *An Evolutionary Theory of Economic Change*, Cambridge, MA: Harvard University Press.

Newell, P., Phillips, J. and Pueyo, A. (2014) 'The Political Economy of Low Carbon Energy in Kenya', *IDS Working Paper* no.445, Brighton: Institute of Development Studies.

Ngigi, A. (2008) Interview, Managing Director, Integral Advisory Ltd, 21 July 2008, Nairobi.

Nordling, L. (2014) 'Africa science plan attacked: Proposed innovation strategy is low on detail and commitments from governments', *Nature*, 510, 452–453.

Normann, H. E. (2015) 'The role of politics in sustainable transitions: The rise and decline of offshore wind in Norway', *Environmental Innovation and Societal Transitions*, 15, 180–193.

Nyaga, E. (2007) Interview, Administrative Assistant, KEREA, 15 November 2007, Nairobi.

Ochieng, F. (1999) 'The amorphous question', *SolarNet Magazine*, 1(1), 19–20.

Ochieng, F., Cohen, A. and van der Plas, R. (1999) *Global Lighting Services for the Poor Phase II: Test Marketing of Small "Solar" Batteries for Rural Electrification Purposes*, Report ESM220/99, Washington, DC: UNDP-World Bank Energy Sector Management Assistance Programme.

Ockwell, D. G. (2008) '"Opening up" policy to reflexive appraisal: a role for Q methodology? A case study of fire management in Cape York, Australia', *Policy Sciences*, 41(4), 263–292.

Ockwell, D. G. and Byrne, R. (2015) 'Improving technology transfer through national systems of innovation: climate relevant innovation-system builders (CRIBs)', *Climate Policy*, 1–19. DOI: 10.1080/14693062.2015.1052958.

Ockwell, D. G., Haum, R., Mallett, A. and Watson, J. (2010a) 'Intellectual property rights and low carbon technology transfer: Conflicting discourses of diffusion and development', *Global Environmental Change*, 20, 729–738.

Ockwell, D. G. and Mallett, A. (eds) (2012) *Low Carbon Technology Transfer: From Rhetoric to Reality*, London: Routledge.

Ockwell, D. G., Sagar, A. and de Coninck, H. (2015) 'Collaborative research and development (R&D) for climate technology transfer and uptake in developing countries: towards a needs driven approach', *Climatic Change*, 131(3), 401–415.

Ockwell, D. G., Watson, J., MacKerron, G., Pal, P. and Yamin, F. (2008) 'Key policy considerations for facilitating low carbon technology transfer to developing countries', *Energy Policy*, 36, 4104–4115.

Ockwell, D. G., Watson, J., Mallett, A., Haum, R., MacKerron, G. and Verbeken, A. (2010b) 'Enhancing developing country access to eco-innovation: The case of technology transfer and climate change in a post-2012 policy framework', *OECD Environment Working Papers*, No. 12, Paris: OECD Publishing.

Ockwell, D. G., Watson, J., Verbeken, A., Mallett, A. and MacKerron, G. (2009) *A Blueprint for Post-2012 Technology Transfer to Developing Countries*, Brighton: Sussex Energy Group.

OECD (1997) *National Systems of Innovation*, Paris: OECD.

OECD (2005) *Oslo Manual: Guidelines for Collecting and Interpreting Innovation Data*, 3rd edn, Paris: OECD Publishing.

Oirere, S. (2012) 'Kenya: Solar financing plan launched', *pv-magazine.com*, 02/05/12. Available at: www.pv-magazine.com/news/details/beitrag/kenya–solar-financing-plan-launched_100 006609/#ixzz2rpuJsqHX [accessed 15 February 2014].

Ondraczek, J. (2013) 'The sun rises in the east (of Africa): A comparison of the development and status of solar energy markets in Kenya and Tanzania', *Energy Policy*, 56, 407–417.

Onyango, C. (2007) Interview, Senior Inspector (Electrical), Ministry of Energy, 13 November 2007, Nairobi.

Osawa, B. (2000) *MESP Project Report*, Nairobi: Energy Alternatives Africa.

Osawa, B. (2008) Interview, Formerly of EAA, Lafarge East Africa (Bamburi Cement), 10 July 2008, Nairobi.

Osawa, B. and Theuri, D. (2001) *Study to Achieve Regional Household Energy Policy Harmonisation – Kenya*, Nairobi: Energy Alternatives Africa and Ministry of Energy.

OSEP (1998) *Business Plan: Orkonerei Solar Energy Project, submitted to the African Development Foundation*, Terat, Tanzania: Orkonerei Solar Energy Project.

Otieno, D. (2007) Interview, Regional Energy Advisor, East Africa, GTZ, 11 December 2007, Nairobi.

Patel, P. and Pavitt, K. (1995) 'Patterns of technological activity: their measurement and interpretation', in Stoneman, P. (ed.) *Handbook of the Economics of Innovation and Technological Change*, Oxford: Blackwell, pp. 14–51.

Peace Corps (1984) *Rural Energy Survey and Profiles: Senegal. A Training Manual*, Peace Corps Information Collection and Exchange, Training Manual T-27.

Perlin, J. (1999) *From Space to Earth: The Story of Solar Electricity*, Ann Arbor, MI: aatec Publications.

Polanyi, M. (1966) *The Tacit Dimension*, London: Routledge and Kegan Paul.

PVGAP (1996) Official launch press release, 25 July.

PVGAP (1998) PVGAP enters implementation phase, press release, 9 February.

PVMTI (2009) Kenya PV Capacity Building Project, webpage of PVMTI announcing capacity-building component of the programme [no longer online].

Radošević, S. (1999) *International Technology Transfer and Catch-up in Economic Development*, Cheltenham: Edward Elgar.

Raven, R. (2005) *Strategic Niche Management for Biomass: A Comparative Study on the Experimental Introduction of Bioenergy Technologies in the Netherlands and Denmark*, Eindhoven: Technische Universiteit Eindhoven.

Raven, R. (2007) 'Niche accumulation and hybridisation strategies in transition processes towards a sustainable energy system: An assessment of differences and pitfalls', *Energy Policy*, 35(4), 2390–2400.

Reinert, E. (2007) *How Rich Countries Got Rich…and Why Poor Countries Stay Poor*, London: Constable & Robinson.

REN21 (2015) *Renewables 2015 Global Status Report*, Paris: Renewable Energy Policy Network for the 21st Century.

Rioba, C. (2008) interview, Managing Director of Solar World (EA) Ltd., 2 March 2008, Nairobi.

Rip, A. and Kemp, R. (1998) 'Technological change', in Rayner, S. and Malone, E. (eds) *Human Choices and Climate Change*, vol. 2: *Resources and Technology*, Columbus, OH: Battelle, pp. 327–399.

Roberts, A. and Ratajczak, A. (1989) *The Introduction of Space Technology Power Systems into Developing Countries, Report No. NASA TM-102042*, Cleveland, OH: NASA-Lewis Research Center.

ROK (1987) *National Energy Policy and Investment Plan*, Nairobi: Ministry of Energy and Regional Development, Republic of Kenya.

ROK (2007) *Physical Infrastructure Sector MTEF Report 2007/8 to 2009/10, Draft*, Nairobi: Republic of Kenya.

ROK (2014) 'KS IEC/TS 62257–62259-5:2013', *The Kenya Gazette*, CXVI(123), 2810.

Rolffs, P., Ockwell, D. and Byrne, R. (2015) 'Beyond technology and finance: Pay-as-you-go sustainable energy access and theories of social change', *Environment and Planning A*, 47(12), 2609–2627.

Romijn, H. and Caniëls, M. C. J. (2011) 'The Jatropha biofuels sector in Tanzania 2005–2009: evolution towards sustainability?', *Research Policy*, 40(4), 618–636.

Romijn, H., Raven, R. and de Visser, I. (2010) 'Biomass energy experiments in rural India: Insights from learning-based development approaches and lessons for Strategic Niche Management', *Environmental Science & Policy*, 13(4), 326–338.

Rose, R. (2004) *Learning from Comparative Public Policy: A Practical Guide*, London: Routledge.

Rothwell, R. (1994) 'Towards the Fifth-generation Innovation Process', *International Marketing Review*, 11(1), 7–31.

Sagar, A. (2009) 'Technology development and transfer to meet climate and developmental challenges', background note for UNDESA Background Paper for Delhi High Level Conference New Delhi, India, 22–23 October 2009, New Delhi.

Sagar, A. and Bloomberg New Energy Finance (2010) *Climate Innovation Centres: A New Way to Foster Climate Technologies in the Developing World?*, Washington, DC: An infoDev publication in collaboration with UNIDO and DFID. Available at: www.infodev.org/infodev-files/resource/InfodevDocuments_1015.pdf [accessed 8 July 2016].

Sagar, A. D., Bremner, C. and Grubb, M. (2009) 'Climate Innovation Centres: A partnership approach to meeting energy and climate challenges', *Natural Resources Forum*, 33, 274–284.

Sahal, D. (1981) 'Alternative conceptions of technology', *Research Policy*, 10, 2–24.

Sanchez, T. (2010) *The Hidden Energy Crisis: How Policies Are Failing the World's Poor*, Rugby: Practical Action Publishing.

Sauter, R. and Watson, J. (2008) *Technology Leapfrogging: A Review of the Evidence*, Brighton: SPRU (Science and Technology Policy Research), University of Sussex.

Schmitz, H., Johnson, O. and Altenburg, T. (2013) *Rent Management: The Heart of Green Industrial Policy,* Brighton: Institute of Development Studies.

Schot, J. and Geels, F. (2007) 'Niches in evolutionary theories of technical change', *Journal of Evolutionary Economics*, 17(5), 605–622.

Schot, J. and Geels, F. W. (2008) 'Strategic niche management and sustainable innovation journeys: theory, findings, research agenda, and policy', *Technology Analysis & Strategic Management*, 20(5), 537–554.

Schot, J., Hoogma, R. and Elzen, B. (1994) 'Strategies for shifting technological systems. The case of the automobile system', *Futures*, 26(10), 1060–1076.

Scoones, I., Leach, M. and Newell, P. (eds) (2015a) *The Politics of Green Transformations*, London: Routledge.

Scoones, I., Newell, P. and Leach, M. (2015b) 'The politics of green transformations', in Scoones, I., Leach, M. and Newell, P. (eds) *The Politics of Green Transformations*, London: Routledge, pp. 1–24.

SE4All (2015) *Progress Toward Sustainable Energy, Global Tracking Framework 2015 Summary Report*, Geneva: Sustainable Energy for All.

Selwyn, B. (2014) *The Global Development Crisis*, Cambridge: Polity.

Sen, A. (1999) *Development as Freedom*, Oxford: Oxford University Press.

Smith, A. (2007) 'Translating sustainabilities between green niches and socio-technical regimes', *Technology Analysis & Strategic Management*, 19(4), 427–450.

Smith, A., Kern, F., Raven, R. and Verhees, B. (2014) 'Spaces for sustainable innovation: Solar photovoltaic electricity in the UK', *Technological Forecasting and Social Change*, 81, 115–130.

Smith, A. and Raven, R. (2010) 'Niche protection in transitions to sustainability', Presentation at the European Association for the Study of Science and Technology Annual Conference, University of Trento, Italy.

Smith, A. and Raven, R. (2012) 'What is protective space? Reconsidering niches in transitions to sustainability', *Research Policy*, 41(6), 1025–1036.

Smith, A. and Stirling, A. (2007) 'Moving outside or inside? Objectification and reflexivity in the governance of socio-technical systems', *Journal of Environmental Policy and Planning*, 8(3–4), 1–23.

Smith, A. and Stirling, A. (2010) 'The politics of social-ecological resilience and sustainable socio-technical transitions', *Ecology and Society*, 15(1), 11.

Smith, A., Voß, J.-P. and Grin, J. (2010) 'Innovation studies and sustainability transitions: The allure of the multi-level perspective and its challenges', *Research Policy*, 39(4), 435–448.

Sokona, Y., Mulugetta, Y. and Gujba, H. (2012) 'Widening energy access in Africa: Towards energy transition', *Energy Policy*, 47, Supplement 1, 3–10.

SolarNet (2001) 'PV Africa pioneer Harold Burris dies', *SolarNet Magazine*, 3(3), 8.

SolarNet (2005) 'Kenya Solar Technician Association (KESTA)', *SolarNet Magazine*, (September) 28.

Sovacool, B. K., D'Agostino, A. L. and Bambawale, M. J. (2011) 'The socio-technical barriers to Solar Home Systems (SHS) in Papua New Guinea: "Choosing pigs, prostitutes, and poker chips over panels"', *Energy Policy*, 39(3), 1532–1542.

Sovacool, B. K. and Drupady, I. M. (2012) *Energy Access, Poverty, and Development: The Governance of Small-Scale Renewable Energy in Developing Asia*, London: Ashgate Publishing.

Starr, M. and Palz, W. (1983) 'Photovoltaic Power for Europe: An Assessment Study', *Solar Energy R&D in the European Community, Series C: Photovoltaic Power Generation, Volume 2, prepared for the Commission of the European Communities, Directorate-General Information Market and Innovation*, Dordrecht: D. Reidel Publishing Company.

Stewart, J. and Hyysalo, S. (2008) 'Intermediaries, users and social learning in technological innovation', *International Journal of Innovation Management*, 12(3), 295–325.

Stirling, A. (2006) 'Precaution, foresight and sustainability: reflection and reflexivity in the governance of science and technology', in Voß, J.-P. and Kemp, R. (eds) *Sustainability and Reflexive Governance*, Cheltenham: Edward Elgar, pp. 335–372.

Stirling, A. (2008) '"Opening up" and "closing down": Power, participation, and pluralism in the social appraisal of technology', *Science, Technology & Human Values*, 33(2), 262–294.

Stirling, A. (2009) 'Direction, distribution, diversity! Pluralising progress in innovation, sustainability and development', *STEPS Working Paper*, 32, Brighton: STEPS Centre.

Stirling, A. (2011) 'Pluralising progress: From integrative transitions to transformative diversity', *Environmental Innovation and Societal Transitions*, 1, 82–88.

Stirling, A. (2014) 'Emancipating transformations: From controlling "the transition" to culturing plural radical progress', *STEPS Working Paper*, 64, Brighton: STEPS Centre.

Stirling, A. (2015a) 'Emancipating transformations: From controlling "the transition" to culturing plural radical progress', in Scoones, I., Leach, M. and Newell, P. (eds) *The Politics of Green Transformations*, London: Routledge, pp. 54–67.

Stirling, A. (2015b) 'Towards innovation democracy? Participation, responsibility and precaution in the politics of science and technology', *STEPS Working Paper*, 78, Brighton: STEPS Centre.

Stua, M. (2013) 'Evidence of the clean development mechanism impact on the Chinese electric power system's low-carbon transition', *Energy Policy*, 62, 1309–1319.

Stuart, B. (2011) 'Ubbink opens East African PV module factory', pv-magazine.com, 31 August 2011. Available at: www.pv-magazine.com/news/details/beitrag/ubbink-opens-east-african-pv-module-factory_100004080/#ixzz2rpsByTma [accessed 15 February 2014].

Swilling, M., Musango, J. and Wakeford, J. (2015) 'Developmental states and sustainability transitions: Prospects of a just transition in South Africa', *Journal of Environmental Policy & Planning*, 1–23. DOI: 10.1080/1523908X.2015.1107716

Tawney, L., Miller, M. and Bazilian, M. (2015) 'Innovation for sustainable energy from a pro-poor perspective', *Climate Policy*, 15(1), 146–162.

Theuri, D. (2008) Interview, Former Acting Head, Renewable Energy Department, Ministry of Energy, 8 July 2008, Nairobi.

Theuri, D. and Hankins, M. (2000) *Kenya: IGAD Household Energy Status Report: Draft Prepared for the 1st IGAD RHEP Kick-Off Workshop*, Nairobi: Department of Renewable Energy, Ministry of Energy and Energy Alternatives Africa.

Theuri, D. and Osawa, B. (2001) *Country Status Report: Kenya, IGAD Regional Household Energy Project No 7 ACP RPR 527*, Nairobi: Energy Alternatives Africa.

Tomlinson, S., Zorlu, P. and Langley, C. (2008) *Innovation and Technology Transfer. Framework for a Global Deal*, London: E3G and Chatham House.

Trainer, T. (2012) 'A critique of Jacobson and Delucchi's proposals for a world renewable energy supply', *Energy Policy*, 44, 476–481.

Turman-Bryant, N., Alstone, P., Gershenson, D. et al. (2015) *The Rise of Solar: Market Evolution of Off-Grid Lighting in Three Kenyan Towns*, Lighting Global.

Turnheim, B. and Geels, F. W. (2013) 'The destabilisation of existing regimes: confronting a multi-dimensional framework with a case study of the British coal industry (1913–1967)', *Research Policy*, 41, 1749–1767.

Ulsrud, K., Winther, T., Palit, D. and Rohracher, H. (2015) 'Village-level solar power in Africa: Accelerating access to electricity services through a socio-technical design in Kenya', *Energy Research & Social Science*, 5(0), 34–44.

Ulsrud, K., Winther, T., Palit, D., Rohracher, H. and Sandgren, J. (2011) 'The Solar Transitions research on solar mini-grids in India: Learning from local cases of innovative socio-technical systems', *Energy for Sustainable Development*, 15(3), 293–303.

UNEP (2006) *Kenya: Integrated assessment of the Energy Policy*, Geneva: United Nations Environment Programme.

UNFCCC (2015) *Joint Annual Report of the Technology Executive Committee and the Climate Technology Centre and Network*, Bonn: UN Framework Convention on Climate Change Secretariat.

Unruh, G. C. (2000) 'Understanding carbon lock-in', *Energy Policy*, 28(12), 817–830.

UON (2014) *Quality Auditors Graduation (ISO 17025)*, 16/09/14, University of Nairobi – Lighting Laboratory website. Available at: http://lightinglab.uonbi.ac.ke/node/3233 [accessed 8 July 2016].

Urama, K. (2014) *Enhancing Adoption and Diffusion of Climate Smart Clean Energy Technologies in Sub-Saharan Africa: Lessons from the Lighting Africa, the Africa Clean Cooking Energy Solutions, and Pro-Poor Low Carbon Development Projects*, Nairobi: ATPS.

Urban, F. and Sumner, A. (2009) 'After 2015: Pro-Poor Low Carbon Development', *IDS In Focus Policy Briefing*, 9, Brighton: Institute for Development Studies.

van der Plas, R. J. and Hankins, M. (1998) 'Solar electricity in Africa: A reality', *Energy Policy*, 26(4), 295–305.

van der Vleuten, F. (2008) Interview, Former Marketing Manager, Free Energy Europe, 17 September, Leusden.

van Eijck, J. and Romijn, H. (2008) 'Prospects for Jatropha biofuels in Tanzania: An analysis with Strategic Niche Management', *Energy Policy*, 36(1), 311–325.

van Lente, H. (1993) *Promising Technology. The Dynamics of Expectations in Technological Developments*, Enschede: Twente University Press.

van Lente, H. (2000) 'Forceful futures: From promise to requirement', in Brown, N., Rappert, B. and Webster, A. (eds) *Contested futures: a sociology of prospective techno-science*, Aldershot: Ashgate Publishing, pp. 211–219.

van Lente, H., Hekkert, M., Smits, R. and van Waveren, B. (2003) 'Roles of systemic intermediaries in transition processes', *International Journal of Innovation Management*, 7(3), 247–279.

van Lente, H., Hekkert, M., Smits, R. and van Waveren, B. (2011) 'Systemic intermediaries and transition processes', in Guy, S., et al. (eds) *Shaping Urban Infrastructures: Intermediaries and the Governance of Socio-Technical Networks*, London: Earthscan, pp. 36–52.

Van Noorden, R. (2014) 'The rechargeable revolution: A better battery', *Nature*, 507, 26–28.

Watitwa, H. (2008) Interview, Chairman, KESTA – Kenya Solar Technician Association, 11 July 2008, Nairobi.

Watkins, A., Papaioannou, T., Mugwagwa, J. and Kale, D. (2015) 'National innovation systems and the intermediary role of industry associations in building institutional capacities for innovation in developing countries: A critical review of the literature', *Research Policy*, 44(8), 1407–1418.

Watson, J., Byrne, R., Morgan Jones, M., Tsang, F., Opazo, J., Fry, C. and Castle-Clarke, S. (2012) 'What are the major barriers to increased use of modern energy services among the world's poorest people and are interventions to overcome these effective?', *CEE Review*, 11–004, Collaboration for Environmental Evidence. Available at: www.environmenta levidence.org/wp-content/uploads/2014/07/CEE11-004.pdf [accessed 8 July 2016].

Watson, J., Byrne, R., Ockwell, D. and Stua, M. (2015) 'Lessons from China: Building technological capabilities for low carbon technology transfer and development', *Climatic Change*, 131(3), 387–399.

WCED (1987) *Our Common Future: World Commission on Environment and Development*, Oxford: Oxford University Press.

Weber, M., Hoogma, R., Lane, B. and Schot, J. (1999) *Experimenting with Sustainable Transport Innovations: A Workbook for Strategic Niche Management*, Enschede: University of Twente Press.

Winther, T. (2008) *The Impact of Electricity: Development, Desires and Dilemmas*, Oxford: Berghahn Books.

WIPO (2011) *World Intellectual Property Report: The Changing Face of Innovation*, Geneva: World Intellectual Property Organization.

Wolfram, C., Shelef, O. and Gertler, P. (2012) 'How will energy demand develop in the developing world?', *Journal of Economic Perspectives*, 26(1), 119–138.

World Bank (2007) *Stand-Alone LED Lighting Systems Quality Screening: Terms of Reference*, Washington, DC: Lighting Africa, 21 November.

Wynne, B. (1993) 'Public uptake of science: A case for institutional reflexivity', *Public Understanding of Science*, 2, 321–337.

Wynne, B. (2002) 'Risk and environment as legitimatory discourses of technology: reflexivity inside out', *Current Sociology*, 50(3), 459–477.

INDEX